Telecom For Dummies®

Cheat Sheet

Information about Your Telecom Carriers

Long Distance Carrier	*Local Carri*
Company Name:	Company N
Account Number:	Account Nu
Customer Service #:	PIC Freeze:
Cust Svc Escalation #:	Customer Service #:
Provisioning Contact:	Provisioning Contact:
Provisioning Phone #:	Provisioning Phone #:
PIC Code:	Features:
10-10 + Dial-around code:	
Date contract expires:	
Intrastate Rate:	
Interstate Rate:	

Information about Your Circuits

Your Dedicated Circuit	*Toll Free Information*	
Circuit ID(s):	Toll Free #:	Ring-to Number
Carrier:		
Circuit Type: T1 or DS3?	DNIS Digits:	
Quantity of Circs:	ANI Delivery: Yes / No	ANI II: Yes / No
Line Coding/Framing:	Overflow Routing: Yes / No	Time of Day Routing: Yes / No
Outpulse Signal/Start:	Geo Routing: Yes / No	Pay phone Block: Yes / No
Trunk Group Config:	Quantity of 8XX:	
Date Circuit Installed:	RespOrg ID Code:	
Trunk Group Name:		

Information about Your Phone System and Hardware

Your Phone System	*Hardware Vendor*
PBX Make:	Company Name:
PBX Model:	Contact Name:
Software Rev:	Contact #:
Channel Bank:	After Hours #:
CSU:	

For Dummies: Bestselling Book Series for Beginners

Telecom For Dummies®

Switched Troubleshooting Essentials (Call Example)

Troubleshooting Question	*Answer*
Origination Phone Number	
Termination Phone Number	
Time & Date of Call	
Call treatment (Less than 24 hours old)	
10-10+ Call made over your long-distance carrier	
10-10+ Call made over another long-distance carrier	
700 Test results = 1-700-555-4141	
Time of Day Specific (Yes or No)?	
Geographic Specific (Yes or No)?	

Dedicated Troubleshooting Essentials (Call Example)

Troubleshooting Question	*Answer*
Origination Phone Number	
Termination Phone Number	
Time & Date of Call	
Call treatment (Less than 24 hours old)	
Rebooted your hardware (Yes or No)?	
Does issue also affect nondedicated calls (Yes or No)?	
Is the circuit released for testing (Yes or No)?	
If no, when?	
If dedicated toll free, do you have a loopback plug for testing (Yes or No)?	
If dedicated toll free, is DNIS correct (Yes or No)?	
If dedicated toll free, is ANI Delivery requested (Yes or No)?	

Item 7085-X.

For more information about Wiley Publishing, call 1-800-762-2974.

For Dummies: Bestselling Book Series for Beginners

by Stephen P. Olejniczak

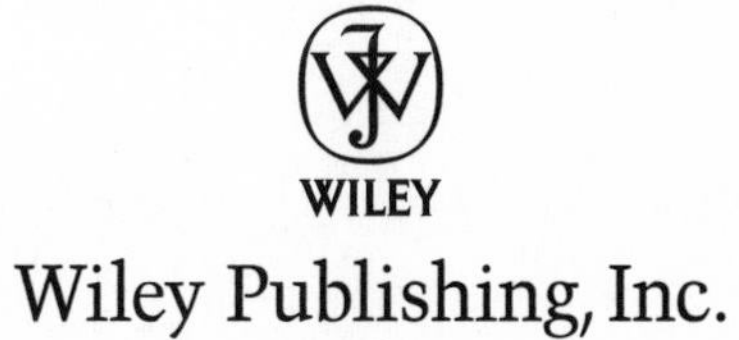

Wiley Publishing, Inc.

Telecom For Dummies®

Published by
Wiley Publishing, Inc.
111 River Street
Hoboken, NJ 07030-5774
www.wiley.com

Published by Wiley Publishing, Inc., Indianapolis, Indiana

Published simultaneously in Canada

For general information on our other products and services, please contact our Customer Care Department within the U.S. at 800-762-2974, outside the U.S. at 317-572-3993, or fax 317-572-4002.

For technical support, please visit www.wiley.com/techsupport.

Wiley also publishes its books in a variety of electronic formats. Some content that appears in print may not be available in electronic books.

Library of Congress Control Number: 2005937358

ISBN-13: 978-0-471-77085-5

ISBN-10: 0-471-77085-X

Manufactured in the United States of America

10 9 8 7 6 5 4 3 2 1

1B/SR/QT/QW/IN

About the Author

Stephen Olejniczak (pronounced ole-en-ee-check) is the Director of Operations for ATI Communications, and has over ten years of telecom experience under his belt. His initial job in telecom was provisioning data services, eventually taking a position as the customer service manager for a small long-distance company, and finally as its manager of dedicated provisioning.

Stephen did not start out in life as a techie, only falling prey to the glamour and easy money after failing to find a career that enabled him to use his Bachelors degree in Cultural Anthropology. He currently lives in the quaint hamlet of Laguna Beach, California, with his wife, Kayley, and a collection of fountain pens.

Dedication

This book is dedicated to the entire telecom industry. From the CEOs of large carriers to everyone that supplies, sells, or uses phone service (I guess that is everyone in the world), I give you this tome of information. The primary group in the industry to whom I dedicate this book are those new employees who have just entered the wild world of telecom. The learning curve in telecom is vertical for at least the first six months, and it is easy to feel overwhelmed. Don't let anyone talk down to you, because we all started out knowing nothing.

Author's Acknowledgments

The greatest motivation and support for this book came from my beautiful wife, Kayley. I am glad that it is complete and we can now travel again.

This book would not be completed if it weren't for the guidance of my Wiley editor, Nicole Haims, who took through the entire process. I must also acknowledge the great work put forward by my technical editor, Frank Piotrowski, who validated everything I wrote, down to the molecular level. Additional props to Kezia Endsley for copyediting assistance. I also received invaluable input from Brady Kirby, of Atlas VoIP Communications, as well as my friends, Chris Lynch and Carl, who kept me on track and running in the data sections. These are only a handful of the brilliant people I have spoken with over the years, from whom I have extracted valuable information that was quickly used to mentor my employees and customers.

Finally, I must give thanks to every customer, salesperson, and coworker who asked me the same questions over and over (and over) again. I wasn't praising you after we chatted at 3:30 a.m. on a Saturday because you wanted to know the country code for Sierra Leone, but now I realize you have given me the depth and breadth of information necessary to write this book.

Publisher's Acknowledgments

We're proud of this book; please send us your comments through our online registration form located at www.dummies.com/register/.

Some of the people who helped bring this book to market include the following:

Acquisitions, Editorial, and Media Development

Project Editor: Nicole Haims

Copy Editor: Kezia Endsley

Acquisitions Editor: Melody Layne

Technical Editor: Frank Piotrowski

Editorial Manager: Jodi Jensen

Media Development Manager: Laura VanWinkle

Editorial Assistant: Amanda Foxworth

Cartoons: Rich Tennant (www.the5thwave.com)

Composition

Project Coordinator: Adrienne Martinez

Layout and Graphics: Carl Byers, Andrea Dahl, Lynsey Osborn

Proofreaders: Leeann Harney, Joe Niesen, Jessica Kramer, TECHBOOKS Production Services

Indexer: TECHBOOKS Production Services

Publishing and Editorial for Technology Dummies

Richard Swadley, Vice President and Executive Group Publisher

Andy Cummings, Vice President and Publisher

Mary Bednarek, Executive Acquisitions Director

Mary C. Corder, Editorial Director

Publishing for Consumer Dummies

Diane Graves Steele, Vice President and Publisher

Joyce Pepple, Acquisitions Director

Composition Services

Gerry Fahey, Vice President of Production Services

Debbie Stailey, Director of Composition Services

Contents at a Glance

Table of Contents

Introduction

Welcome to *Telecom For Dummies,* a book for people who work in telecom (99 percent of whom come into the industry through no fault of their own). You're probably a very smart person, and so your boss decided to give you the responsibility of handling that expensive communication network that keeps the company in business. Don't worry! This book can help you work through almost any question you have about telecom. In the end, you will be very comfortable with your new environment and you will continue to impress others as the wonder kid they always believed you to be.

This book contains everything you need to know to order, maintain, and troubleshoot basic phone service. It covers the nuts and bolts of how phone systems work, why they work, and why it sometimes takes so long for them to work. When you have questions, simply track down the chapter and subsection that covers the issue in question, and after a little reading, you will be able to talk to any technician with confidence.

About This Book

This book was not intended for bedtime reading from cover to cover. It is a very helpful reference for telecom products, applications, and troubleshooting. The first few parts cover finding a phone service that best suits your business needs. Another part provides the ins and outs of ordering what you need. If you already have a phone system set up, move to the part that covers what you need to know to troubleshoot the circuits and systems you've installed.

Every chapter has been written with you, not an MIT technician, in mind. The information is easy to understand and digest, even if you have absolutely no prior telecom knowledge. If additional information might be helpful, I refer you to another chapter for more information.

Telecom For Dummies is applicable to almost all phone service in North America, including Canada, many of the Caribbean countries and Guam. The regulations and infrastructure for telecom vary between most countries, and although some aspects may be applicable in Europe and Asia, the steps for ordering and testing systems vary.

Conventions Used in This Book

We've used a few conventions in this book to make it easier for you to spot special information. Here are those conventions:

- New terms are identified by using *italic* and followed by a short definition.
- Web site addresses (URLs) are designated by using a `monospace` font.
- If I tell you to dial a number or type a specific command, the command appears in **boldface.**

What You Don't Have to Read

You don't have to read anything that doesn't apply to your needs. If you don't have a phone system, or dedicated circuits, or place any international calls, for example, you can ignore the sections that cover them. The book contains enough great information that you won't hurt my feelings by jumping from chapter to chapter (or even from section to section).

Icons Used in This Book

Telecom For Dummies includes icons that point out special information. Here are the icons I use and what they mean:

This icon makes you feel like a real telecom pro. It highlights special tricks and shortcuts that make understanding and maneuvering within the vast telecom world even easier. Don't skip this information!

This icon reminds you of important information that can be far too easy to forget and which can cause a lot of frustration when you do forget.

Be careful when you see this icon. It points out an area where you'll want to be extra cautious so that you don't cause yourself problems. It also tells you how to avoid the problems.

Technical Stuff is information for folks who want to know all the geeky details.

Foolish Assumptions

I assume that you have seen a phone, dialed a phone, and have had a conversation on a phone before. In addition to that, your job is somehow linked to buying, selling, using, or supporting some kind of telecommunications service. I assume the following about your everyday contact with telecom tools and systems (perhaps you don't fit in every one of these scenarios, but you recognize yourself in at least a few of them):

- You have to make decisions on buying or upgrading phone services.
- You have had problems ordering phone service in the past and want to know some tips on how to keep moving forward without unnecessary delays.
- You have an inventory of toll-free numbers that you must manage.
- You want to find the most efficient way to speak to your carriers and hardware vendors so they understand your needs and expectations.
- You would like to have the power to troubleshoot issues, such as failed calls and quality issues, without relying on someone else for answers.

How This Book Is Organized

Telecom For Dummies has six parts. Each part is self-contained, but all the content is somewhat interconnected. That way you'll see the most useful information without a lot of boring repetition.

Part I: The ABCs of Telecom Services

This part explains the landscape of telecom, the key players, and how they work together. I describe the differences in responsibilities between local, long-distance, and wireless carriers. I also include information that introduces the basic telecom features and options.

Part II: Reviewing Telecom Products and Prices

Not every telecom product is right for every customer. Part II reviews the most common telecom products so that you can evaluate which of them are right for your business. This part covers a wide range of services, and helps you analyze whether you should jump from regular (switched) phone service to dedicated phone service. It also gives you the lay of the toll-free land and helps you maneuver through your phone bill, looking for areas that are costing you more money than they should. Stop the bleeding in this part, and figure out which of your potential telecom investments will give you the best return.

Part III: Ordering and Setting Up Telecom Service

The second most painful aspect of telecom is ordering new service (see Part IV for the most painful aspect of telecom). This part guides you through the ordering process for all services, from regular (switched) phone lines to dedicated circuits, to toll-free service. Because dedicated and toll-free services are complicated, I include a chapter in this part that goes the extra mile, showing you how to activate these services after you order them. All along the way, I tell you about potential pitfalls so that you can successfully avoid them.

Part IV: Taking Care of Your Telecom System

The most painful aspect of telecom is troubleshooting problems. The issue afflicting your system may be huge or microscopic, but you still need to fix it. Part IV covers troubleshooting switched phone lines, dedicated phone lines, and toll-free service in a step-by-step manner that enables you to make quick work of almost any problem. By following the rules I set out for you in this part, you can systematically identify problems and keep your technicians from going on a wild goose chase.

Part V: What's Hot (Or Just Geeky) in the Telecom World

The chapters in Part V cover the world of telecom — beyond voice phone calls. I cover the basics of data transfer technologies, and the hottest buzzword in telecom right now, VoIP. Part V won't show you how to write the code to transfer the data, but it does give you an overview of the newest and greatest technology, gives you some hints on pricing, and tells you about the hardware required to create a data-transfer interface with your carrier.

Part VI: The Part of Tens

Part VI covers industry buzzwords, personality disorders, and where to go for help. How's that for a mélange? When you use common acronyms casually when speaking to your carrier reps and technicians, they know that you're not clueless about telecom. The section on telecom traits provides behavior to look out for and hints on how to avoid it. Finally, the last chapter lists resources to tap into when you are at your wit's end.

At the end of the Part of Tens is an appendix I tacked on to show you how to make male and female loopback plugs. These little gadgets are simple and small, but they are invaluable to troubleshooting phone systems.

Where to Go from Here

The best place to start on this book is the Table of Contents or Index if you want to key in on information about a specific topic. Not sure whether you have a dedicated circuit or what your local carrier's responsibilities are? Chapter 1 can help you get your bearings.

Depending on the specific aspect of telecom you need to research, you may want to jump to Chapter 6 to find out about what's going on with your phone bill, or if you're experiencing a major malfunction right now, skip to Chapter 13 to figure out how to troubleshoot your dedicated circuit. This is your buffet of telecom goodies; check out the entire offering and dive into the sections you think are tasty.

Part I
The ABCs of Telecom Service

"It's all here, Warden. Routers, hubs, switches, all pieced together from scraps found in the machine shop. I guess the prospect of unregulated telecommunications was just too sweet to pass up."

In this part . . .

You get an overview of the key players in telecom, as well as an introduction to the structures that allow these players to work together. This part also offers you guidelines to identify the phone system you are currently using, to help you determine whether your business is a good candidate for dedicated phone service, and to help you identify the hazards of telephone fraud. This part wraps up with an overview of how calls are classified and explains why your international call to Canada doesn't require you to dial 011.

Chapter 1

A Buyer's Scoop on Telecom

In This Chapter

- Getting a general understanding of telecom
- Checking out the players
- Planning and implementing effective telecom systems
- Making major telecom decisions
- Finding people to help you make the right choices
- Setting up for future expansion
- Troubleshooting your telecom system

Every business in the world either uses or could benefit from using phone service. Even if your business deals directly with the public, you still need to call your manufacturer for updates on shipments, check on the ads you have going out, and make sure your accountant has all of your financial information. As your business grows, so does your need for phone lines, not just to enable you to speak to your customers and suppliers, but also to transmit data between offices, and possibly even maintain a Web site to receive and track orders.

Growth in business is great, but you may easily find yourself overwhelmed by the logistics involved in all the changes. The phone system you had 2 years ago may have been perfect for you at the time, but today you have 20 new employees in house, 5 new outside sales reps, and the Web site for Internet orders is being released next week. If that is the case, you have probably outgrown your old telephone system. The important thing to know is how to gauge the technology you have, where you are going, and what technology you need to get there. In this chapter, I help you figure out what you're dealing with and get you started on your quest to improve your telecommunication, or *telecom* system.

Assessing Your Telecom Services As They Are Now

At the most basic level, *telecom* refers to any service that is provided over phone lines. Included in this definition are

- Internet connections over *dedicated circuits* or dial-up modems
- Regular phone lines used to place and receive calls to your mom
- *High-speed data links* used for transmitting information

When you boil things down, all phone service can be reduced to the same basic principles of transmitting and receiving signals over fiberoptic or copper wire.

For example, if a backhoe cuts through a fiberoptic cable that supplies your Internet connection, along with the service from your local carrier, your only connection to the outside world becomes your mobile phone. If your mobile phone provider also uses that link to complete calls, you still may not have a connection to the outside world. With the increase in *redundancy* being built into the local and long-distance networks, you shouldn't have an outage that completely shuts down your service. If you have a good telecom system, when (not if) a breakdown in the network occurs, stopping calls from passing through the east side of town, your system can automatically route calls to an alternative network on the west side.

Many companies rely heavily upon their phone service, and any outage, no matter how small, has a large impact. Sometimes the problem is an act of God, (say your carrier's switch is struck by a bolt of lightening, reducing it to a smoldering brick of silicon and steel valuable only as a piece of modern art). Other issues may be the result of bottlenecks in your own phone system, or just a need for more phone lines. The first step to taking care of these issues is to assess the phone service you need right now.

Identifying Your Carriers

A telecom *carrier* is an entity that provides and bills for phone service. It would be convenient if I could classify a carrier based on the fact that it owns a large fiberoptic network and multimillion-dollar phone switches, but not all of them do. Some carriers own no hardware and simply contract with other companies that have sophisticated hardware networks.

Actually, all carriers have contracts with other carriers to sublet space on their networks. Subletting enables them to build more redundancy into their systems (which is good thing for customers). In some areas, subcontracting also helps carriers get substantial price breaks when they try to negotiate new contracts to gain entrance into a specific market.

In fact, the best way to understand carriers is based on their functions. Carriers treat local networks, long-distance networks, mobile networks, and more.

Looking locally

Local carriers provide local service. If you call from your office to the building next door, your local carrier receives the call and completes it to the other building. In addition to providing local calling services, local carriers also provide you with dial tone on your residential or simple business phone lines, assign your phone number, provide 911 (emergency service), 411 (information), toll-free service and a host of other features like call waiting, caller ID, and sometimes even voicemail. One of the most important functions of the local carrier is to identify every call you make as being local, long distance, or toll free, and then to route it to the appropriate carrier to complete.

The industry refers to local carriers primarily as *local exchange carriers* (LECs). When you think of LECs, names like Bell Atlantic, Bell South, Ameritech, or Verizon come to mind. These are all companies that were lucky enough to be given the limited monopoly to provide local service to a specific geographic area. America was carved up based on local population statistics and state borders when the Ma Bell monopoly got broken up in the 1980s.

LECs are also known by other names. The LECs that were part of the initial monopolies given by the U.S. government and generally have the word *Bell* in their names are sometimes grouped as *RBOCs* (pronounced *ahrr-boks*). An RBOC is a *Regional Bell Operating Company.* At times, RBOCs are also referred to as *ILECs* (pronounced *eye-leks*). The ILEC designation identifies a carrier as being the senior LEC in the area, specifically, the *incumbent local exchange carrier.* Throughout this book, I avoid using too much jargon by simply referring to the carrier that supplies your local telecom service as your local carrier.

Introducing competing local exchange carriers

A *competing local exchange carrier* (CLEC) does the same work as other local carriers. The special name simply signifies that these carriers arrived on the scene later than the baby Bells referred to in the previous section, "Looking locally."

Understanding how CLECs got a piece of the pie

There were two basic ways the CLECs gained the ability to provide local service to their customers. They either spent millions of dollars on cabling and hardware to replace the networks of the ILECS or they contracted with the ILECs to resell ILEC services at discounted rates. Companies like ICG Communications and XO Communications spent millions of dollars building their infrastructures and securing contracts in the local area to allow them to compete with the ILECs.

All this work was done in an attempt to gain access to the businesses and residences in the CLEC's target market. The advent of functional Voice over IP (VoIP) enables CLECs to use the existing copper wires that connect networks to the Internet so that they can also carry phone calls. Making phone calls online greatly increases the number of potential customers for the CLECs and reduces the overhead required to connect to them. I discuss VoIP in Chapter 15.

Any company that can assign you a phone number, but was not given the monopoly for the area by the government, is technically a CLEC. This category includes wireless providers, *VoIP (Voice over IP)* providers, and the true CLECs whose sole goal, in the spirit of free-market economics, is to compete with (and eventually replace) the Verizons and Ameritechs of America.

Going for long distance

Long-distance carriers receive and complete calls that terminate outside of the U.S., across a state border, or across a geographically defined border within the state called a LATA *(Local Access Transport Area)*. Long-distance functions were specifically denied to the local carriers during deregulation and spawned the growth of the long-distance industry.

You can break long-distance carriers into two categories:

- **Facilities-based long-distance carriers:** The first companies that come to mind when you think of long-distance carriers are AT&T, Sprint, and MCI. These are companies that own huge telecom networks, million-dollar telephony routing switches, and enough cabling to tie us to the moon ten times. All the equipment and stuff — the hardware and cabling — that these companies own establish these carriers as facilities-based providers. Generally, only large companies worth hundreds of millions of dollars fall into this category, but ever since the long-distance industry was born, the number of facilities-based long-distance carriers has been growing.
- **Switchless resellers:** Along with the AT&Ts and MCIs of the U.S. came a new breed of long-distance carrier. These companies don't own any

hardware or network facilities, but simply resell existing services from the larger facilities-based carriers like Sprint or Qwest. So-called *switchless resellers* sign contracts with large carriers for a specific per-minute rate and then resell the service to companies and residences for a profit.

Generally, switchless resellers are much smaller companies and fit a niche market of customers who want to mix the personal service typically available from a smaller company with the stability and network functions that only a large company can provide. In return for an inexpensive per-minute rate from the carrier, the switchless reseller takes over the job of providing all customer service functions and billing services to customers. Switchless resellers can be your business's best friend, *and* save you quite a bit of money; or they can be a nightmare and cost you a lot. If you're considering using a switchless reseller, you should research the company and its management team. Ask for some references before you jump into a contract with any switchless reseller.

Working with wireless providers

Wireless communication is a method of transporting a call more than it is a standard division of labor within the telecom world. The wireless industry was born after the telecom industry was deregulated and as a result, the industry enjoys all the benefits of the breakup. The wireless companies function just as CLECs do, but can also provide long-distance service.

Wireless providers have the benefit of much lower start-up costs than other telecom providers, because they simply install hardware to transmit and receive wireless calls where the hardware is needed instead of digging up endless miles of roads to lay down new cabling. Today, technology has evolved to the point that you can send and receive e-mail and text messages, surf the Internet, and download video — all with your wireless phone. This is the one part of telecom that enables you to have it all in one device.

Knowing Why Your Company Needs Telecom

The first question you need to ask yourself is, "What do I want telecom to do for me?" The answer to this question depends on several factors, including the following:

- Your industry
- The type of business you have
- The size of your business

If your business is in the car rental industry and all of your competitors have toll-free numbers, then you probably need to keep up with the Joneses. If your business deals only with exotic car rentals, a toll-free number may not be mandatory, but a Web site displaying the Porsches and Ferraris may be a must. A national car rental chain may need an extensive data network to keep everyone informed about which cars are available at all locations. Take a moment to look at your company and your competitors to find out where you want to be in your industry. The services you and your customers demand will fall into five categories:

- **Voice service:** This is standard phone service that you have been using since the first time you called your Grandma to thank her for a birthday present. You need voice service (and you probably already have it), so a bigger question is whether most of your calls are outbound, inbound, or a relatively even combination of outbound and inbound calls. If most of your calls are outbound, and you only receive a few inbound calls a week, then standard voice service may be all your business needs.
- **Toll-free service:** A *toll-free number* is a special number that enables calls to come in from anywhere without the caller incurring a charge. The called party pays all the charges for the call. How much the called party pays for a call depends on the arrangements made with the long-distance carrier. If you look at all of your voice calls and find that the majority of them are inbound from customers, you could probably benefit from toll-free service. For many companies, not having a toll-free number is disastrous. If your car breaks down and you have to call for assistance from a pay phone, would you rather pay 50 cents to contact the answering machine of a tow truck or taxi company, or would you rather call for free? If you're on hold for a couple of minutes on a toll-free call, at least you're not paying for it. Toll-free service is more complex than regular outbound service, and is covered in detail in Chapter 5.

  People prefer to dial 800 numbers when they are going to order something, be it Diamonique jewelry or towing service. If you have to pay $75 to tow your car (and goodness knows how much to repair it), at least you can rest easy knowing that your phone call to the towing agency is the only part of the experience that doesn't cost a thing.
- **Data services:** The term *data services* usually refers to non-voice communications such as e-mail, text messaging, and fax services that are transmitted over a private network. In every company, information needs to travel from one side of the office to another side of the office. The sales team closes the order and relays it to the order-entry department. The order-entry department logs the order and sends it to the manufacturing department; the manufacturing department builds and tracks the order; and the shipping department sends the order to the customer. For smaller companies, all these steps may be taken care of

with an Excel spreadsheet or an industry-specific software package. As your company grows, however, more people in more offices need the same information, and the physical location of the offices is farther and farther apart. Last year, your business's second location may have been across the hall in the same building, but today you could have a London office that generates information that comes to you in Los Angeles before you send it off to your Tokyo division. Data transmission can be done in many ways, and your decision will be based on the physical location of your offices, the frequency of data transmissions, and the amount of data being transmitted. The up-and-coming data service technologies, as well as the old standbys, are covered in greater detail in Chapter 14.

- **Internet Service:** Every office needs a connection to the Internet, if for no other reason than to enable employees to send and receive e-mail and while away the hours Googling their names. Many companies use the Internet to research their competitors, or they may use the Internet to visit suppliers' Web sites to place orders, review shipments, and request services. I can safely say that this portion of your telecom needs will not decrease in the future — it will increase incrementally. Many companies integrate the private networks transmitting their data services and their need to access the Internet on the public network.
- **Wireless service:** Wireless service can encompass all the other categories in this list, but because wireless service is usually provided by a different company than the one that provides your land lines, you need to view it as a separate telecom service. If you have an external sales force or a 24-hour service department that makes house calls, wireless service is an essential part of your business.

Introducing Dedicated Long-Distance Circuits

There is one big hurdle in telecom, and after you jump this barrier you move from Junior Varsity to Varsity. That transition is the jump from regular phone lines (what the industry calls *switched* lines) for long-distance use to *dedicated long-distance circuits.*

You probably use regular *switched* phone lines at home every day. Many small businesses have them, and even large businesses have a handful of switched lines for backup purposes, to run security services, or to handle overflow phone calls. Switched lines are wonderful because you can do everything with them: You can call 411 or 911, dial a toll-free number, call your brother next door, or your aunt on vacation in Istanbul. What is even better is that other people can easily call you.

Getting dedicated locally

Dedicated circuits are classified by the carrier to which they terminate. A dedicated circuit that ends at your long-distance carrier is a *dedicated long-distance circuit.* A dedicated circuit that ends at your local carrier is a *dedicated local circuit.* Local circuits can be filled with phone numbers that act just like switched lines.

Those businesses with dedicated local services have several phone numbers for each local circuit, probably 24 (one for each usable channel). The only problem you may have with dedicated local circuits is that your long-distance carrier sees the outbound calls coming from your local circuit as originating from a phone number and not a circuit. Your dedicated long-distance rates are only applied on calls that originate on a dedicated long-distance circuit. Calls originating from a local circuit appear as coming from a regular switched phone line, and as a result, your long-distance carrier rates these calls at a higher switched rate. There are instances where the benefits of a local circuit outweigh the increased per-minute cost you will see from your long-distance carrier, but you should keep the financial consequences in mind when you decide to make the leap to a dedicated local circuit.

Dedicated long-distance circuits are good *only* for long-distance calls, whether they are to another state or another country. However, dedicated long-distance circuits do have some downsides. Several of the features that you take for granted are not provided on them. Services that are not available on dedicated long-distance circuits are:

- Access to 911 emergency service.
- Access to 411 information service.
- Access to 611 telephone repair service.
- The ability to dial toll-free numbers.
- The ability to dial long-distance bypass codes for your calls. These are the 10-10 codes you may have heard about on TV.
- The ability to receive calls by any means other than a toll-free number.

Actually, no dedicated circuit has a phone number naturally built into it. You can receive inbound calls on your dedicated circuit, but only with a toll-free number pointed to your circuit or special *Direct-Inward Dialing* (DID) service from a carrier. DIDs only allow you to receive calls, and act like toll-free numbers, sending calls into your dedicated circuit, but not being listed on your caller ID when you dial out from it. DIDs are generally only available on circuits set up by your local carrier, and not for long-distance circuits.

With all these limitations on long-distance circuits, you might wonder why anyone orders them. The reasons add up quickly in per-minute cost savings. Just as you get a better deal when you buy anything in bulk, you receive a better per-minute rate on a dedicated circuit.

If you're pushing enough long-distance minutes through your existing phone lines, you simply have to add up the numbers to see whether jumping to a dedicated circuit will benefit your business in the short and long term.

The leap from switched service to dedicated circuits is the only significant jump to be made in telecom. The industry recognizes two categories of pricing: one for switched phone lines and the second for service provided via dedicated circuits. Unless you elect to become a carrier and become eligible for *dedicated carrier pricing*, there isn't another plateau in the industry available capable of reducing your per-minute cost by 40 percent. The moment you install additional dedicated circuits you are not guaranteed a better rate, but you do have more leverage when you renegotiate your contract. The cost justification for dedicated service is covered in detail in Chapter 2 with some real-world numbers that can guide you to see if it is right for your business.

Finding People to Help You Make the Right Choices

The one thing I can guarantee about telecom is that it is always changing. The hot technology of today will be replaced by the hot technology of six months from now. Unless you work in the telecom industry, it is very difficult to keep up on the changes and figure out how your company can best use them. You need a support team. In telecom, just as in life, everyone is a potential source of information. Keep that in mind when it comes to your phone service. Some of the people to whom you have access could save you thousands or millions of dollars on your phone bill over the course of a year. The following sections point out the important people you need to know and trust if you want to use the best technology at the lowest rates.

Meeting the sales rep

Every carrier has a sales force whose members would love to meet you if they haven't already. The sales people come in two flavors, the *carrier sales rep,* and the *independent sales agent.* They are distinctly separate creatures and should be treated as such.

- **The carrier sales rep** works solely for, and is paid by, one carrier (whether a long-distance carrier, a local carrier, a wireless service provider, or another carrier). Whatever your application is, the rep finds some solution, based on the technology the carrier has available. Remember, this solution may not be the best solution for *you,* but it is the best solution the carrier can offer. For example, maybe you need a low-cost calling option for your sales force in Germany. Instead of offering you international toll-free numbers, the carrier sales rep may offer calling cards with international origination.
- **The independent sales agent** is generally a better person to chat with. She probably represents four or five different carriers, all with different strengths, weaknesses, and support levels. The best thing about an independent sales agent is that the agent doesn't have any allegiance to any one of the networks. The agent will sell you toll-free numbers from MCI just as quickly as she'll offer outbound dedicated circuits from Qwest and calling cards from Broadwing. The independent sales agent really doesn't care which options you choose, just as long as you're happy and you renew your contracts.

If you're not sure what you want to do, talk to your carrier sales rep first. This person is outstanding if you're fishing for new ideas. Start your conversation by asking the rep to go over your existing telecom services; then discuss what you want telecom to do for you. Ask what alternatives the carrier has for accomplishing those goals. I promise that if you present a salesperson with a telecom problem, he or she will come back with at least one solution, even if it's not the best one. If you don't have an existing long-distance carrier sales rep, you can have one assigned to you by calling the customer service department and asking to speak to someone.

Seeking out a hardware vendor

If you have anything more complicated than single-line local carrier services, you need a hardware vendor. If you have a phone system, you may already have a hardware vendor that you call when you need service.

The technician that actually services your hardware probably loves all things complex and convoluted. Reps and techies for hardware vendors are outstanding individuals to chat with when you are about to make any change to your phone service. Your vendor rep or technician probably knows the latest technology on the market and would love to tell you about it — you may even get more information than you ever asked for or wanted. The best thing is that your hardware vendor already knows what kind of phone system you have — and whether the latest gadget can actually be installed with it.

It is wonderful that VoIP is the darling of the telecom world today, but if you have to completely rip out $35,000 in phones and equipment and shell out $100,000 for new hardware, this technology may not be such a great deal after all. Your hardware vendor can alert you of this fact immediately.

Begin your quest for new and better solutions with a carrier sales rep or independent sales agent, and then circle back to your hardware vendor to confirm that your plans won't require a complete overhaul of your existing system. If you don't have a vendor, check out the following sections, "Finding a hardware vendor from your sales agent" and "Finding a hardware vendor from the manufacturer."

Finding a hardware vendor from your sales agent

If you inherited your phone system, you may not know where it was purchased. Likewise, if you don't have any phone system to speak of, you need to find a vendor on your own.

You're better off finding a vendor today than waiting until your system goes down or you're ready for an upgrade. Finding a new vendor who can give you an accurate assessment of your current telecom system can take three months or more. Your sales agent is generally a good person to ask for references on hardware vendors. Over the course of a career in telecom, sales agents bump into good vendors and bad vendors. Your agent can happily refer you to a competent and professional company, often offering testimonials about how this or that person helped the agent or a colleague out of a tight spot.

Ask for at least two different companies, just so you have some options. Hardware vendors generally specialize on certain makes or models of specific phone systems, so your first question should be, "Which phone systems are your specialties?" If a vendor only handles Newbridge channel banks and you have an Avaya system, you need to keep searching. Honest vendors have no problem referring you to a business in the industry that specializes in the hardware you have.

Finding a hardware vendor from the manufacturer

If you strike out with your sales agent and their referrals (see "Finding a hardware vendor from your sales agent"), don't give up; contact the manufacturer of your hardware.

Not sure who made your hardware? Take a small walk to your phone closet. No idea where your phone closet is? Follow these steps to find out more about your system and use the info to track down a vendor:

1. **Track all of your phone lines to the place in your office where they converge into one large plastic or metal box.**

That's your phone closet. On the outside of the box, you see a manufacturer name and model. For example, the outside of the box might say something like `ADTRAN TSU600` or `Newbridge 3624 Mainstreet.`

2. **Write down the names on all the boxes your phone lines go through before they leave the phone closet.**

   Be especially careful to note anything that has the words *MUX, multiplexer, PBX,* or *key system.* The more information you have the better.

3. **After you acquire information about the manufacturer, track down the companies on the Internet.**

   Somewhere on the manufacturer's Web site you will see a section for service. Many manufacturers even include vendor locators.

4. **Track down two or three of the nearest service companies and ask to conduct a phone interview.**

   You are looking for a rep that is professional, responsive, and knowledgeable. If you're not filled with confidence when you chat with a rep, you probably won't be filled with much confidence when you see the company's techs stumbling through your equipment.

When you've settled on a vendor, you need to have a rep come to your site to give your hardware a once-over. This meeting gets the vendor familiar with your system — and you familiar with the technicians.

Use this little meet and greet to have the talk about how your system is serving your company's needs. The manufacturer may have released a new card that gives your system twice its current capacity, three times the current number of features, and costs half what you would spend to repair the system in the event of its impending meltdown. You won't know this information until you ask — this is definitely a conversation worth having. See the next section, "Planning for Growth," for more information.

Planning for Growth

No company stands still. Your business is either expanding or (I hope not) contracting. If your company is bursting at the seams, you need to begin chatting with both your hardware vendor and your carrier sales agent today. To service the new employees, new departments, and changing requirements of the company, you may need to upgrade your phone system, add more phone lines, install dedicated circuits, or possibly start over with a brand-new phone system (if your current system has no room left for expansion).

When you're ready to make changes to your system, and after you've established a relationship with your carrier sales agent or independent sales rep, as well as with your hardware vendor, you need to bring everyone together at your office for a chat so that you can determine the best strategy for changing your phone service to respond to your business's growth.

Having a conversation with your hardware vendor is very simple. Your questions are:

- How many more telephones can I install on my system as it is right now?
- How many more lines can my system handle from the carrier as it is right now?
- If I buy additional cards for my current phone system, how many phones for employees can I add before I max out the system?
- How long does it take to receive and install the new cards?
- How much do cards cost to buy and install? Are there any other fees I should know about?
- What are my options when I have no more capacity on the existing phone system?

Getting answers to these questions gives you a sense for the system's current limitations, a rough timeline for adding capacity to your system, and the general cost for the various upgrade options.

On the carrier side, the questions are even fewer. If you are adding individual phone lines, you need to contact your local carrier to find out the installation and monthly cost to add lines, as well as the standard interval to have the lines brought to your phone room and installed. If you're adding dedicated service from your long-distance carrier, your questions are essentially the same. The basic information you need has to do with

- Timelines for installing new circuits.
- Charges, both *monthly recurring charges* (MRCs) and installation charges and other one-time costs (also known as *nonrecurring charges* — NRCs).

When you have all the information you need, you can plan accordingly so that expansion is methodical and planned, as opposed to being anxiety ridden and reactionary. The specific charges associated with dedicated circuits are covered in Chapter 2.

Troubleshooting All Things Telecom

There are three groups of telecom people you need to have on your speed dial *before* you have a crisis. These are the individuals who troubleshoot and repair your phone system problems when you cannot dial out, have static on the line, are dealing with echo (echo, echo), or any issue that prevents you from completing a call. Although I offer extensive troubleshooting information in this book that you can use to isolate problems, you still need the right company to correct problems. These are not DIY tasks. Be prepared to call the following individuals:

- **The customer care center for your long-distance carrier(s):** These people can help you troubleshoot and repair anything that is the responsibility of your long-distance carrier. Not every problem belongs to your long-distance carrier, but when you identify a long-distance telecom issue, you need to call the long-distance carrier. In the event of a serious problem, keep a full phone list that escalates all the way to the vice-president level. Nothing feels worse than having a contact number for the lowest end of the support spectrum, and no way to escalate.
- **The trouble reporting center for your local carrier:** Generally, you can reach this office by dialing 611 from your phone, but if your phone lines are completely out, you may have to use a cellphone. Unfortunately, dialing 611 from your cellphone won't do you any good at all. Instead, find the customer service number on an old phone bill or search the Web for the customer service number, and keep the number handy. As with your long-distance carrier, have an escalating list of phone numbers so that if you don't receive immediate resolution you can talk to someone with more authority.
- **The service number for your hardware vendor:** Keep the number of your hardware vendor close at hand. Sometimes phone issues are hardware related, and in those situations, your long-distance or local carrier can do nothing to resolve them.

Chapter 2

Making (And Living with) Telecom Decisions

In This Chapter

- Determining how many phone lines you need
- Finding your phone system
- Identifying and evaluating the system
- Understanding the cost of going dedicated
- Preventing fraud
- Negotiating the best telecom deal for you

The good news about the world of telecom is that it is always changing. The other good news about telecom is that it stays the same. You can buy a cellphone right now that can surf the Internet, access your e-mail, and send pictures and video. You can also buy a vintage rotary-dial phone from 1935, plug it into the wall jack in your house, and it will work just as well as the day it was built. So even though telecom is in a constant state of expansion, the continuity enables almost every era of the technology to remain in use today. As new services are released, they fit within the existing infrastructure, but never replace it.

An old phone system is not necessarily worthless, and a new system is not always mandatory. The changes you make to your telecom system depend on your company's needs. As carriers roll out more services, you may find yourself overwhelmed by all the hardware and service choices that appear on your invoice, and the basic tenets of how to decide what to do next. Don't worry. This chapter gives you an understanding to deal with all of it.

Analyzing How Many Phone Lines You Need

Your decision about the number of phone lines your business needs will vary based on the specific profile of your company. Don't be snowed into believing that you need one line for every employee. On the other hand, you shouldn't cut corners. Here are a couple of scenarios to help you understand what you should consider in your decision:

- **Tons of employees in the field:** Say you run a construction company with 50 employees. Four of your employees work in the office and the rest of the crew is out on job sites. You don't need 50 phone lines. You might purchase four lines for your in-house employees, one spare line for employees to pick up voicemail, and a line or two for your fax machines and be fine. You can noodle around with the configuration if you want. Maybe your construction company is very busy. You may need three or four lines for faxes.
- **Tons of calls coming in and going out all day:** You own a telemarketing company with 25 people who make calls 8 hours a day. Your employees also need to connect customers to a conference call with a verification service. You may need 35 or 40 lines just for the calls your employees make, and then 2 lines for faxes.

As a basic rule, you should have one phone line for every employee who makes calls in your office — up to about 15 employees. If you have more than 15 employees, the least expensive option is to jump to a dedicated circuit that contains 24 phone lines. See the section later in this chapter, called, "Deciding whether to Get Dedicated or Stay Switched."

If your business is growing, here's an interesting fact. You will need fewer lines per employee as you increase the number of employees in a standard office environment, not because employees will make and receive fewer calls (you can expect that they'll probably make more calls), but because the statistical probability that everyone will be on the phone at the exact same time is pretty slim. You face a greater likelihood that one person will be hanging up and doing work as someone else picks up the phone to start another call.

To get a handle on your business's actual call volume, you *could* pore over the itemized list of your phone calls to calculate the total number of calls active during peak times. Have fun — *not!* A more realistic exercise is to scan your phone bill and determine whether every phone number is being used. If you have several phone numbers that are not designated as inbound faxes or are lines for your security system and they have no usage, you probably have too many lines.

On the other side of the equation, if anyone (customers, employees, associates) ever complains of hearing a recording from the phone system that says that the "service is unavailable," you probably need to open some more lines.

Before you run out and buy a bunch of new phone lines, first ensure that the bottleneck in your phone system isn't caused by another problem (maybe a card failed and you lost ten lines in your phone system because of it).

After validating your hardware, order additional lines to match the number of people affected by the shortage of lines. The Human Resources department may have hired on 20 more people you didn't even know about. If 10 people in your office can't get outside lines during peak times, order 12 more lines.

Locating Your Phone System: What's in Your Closet?

Before you can maintain, upgrade, or replace your phone system, you need to find it. If you are in charge of the phones in your office, take a field trip to get to know your phone system. Or talk to the hardware vendor that takes care of your phone system and find out where the system lives.

Phone systems are generally shoehorned into a small broom closet somewhere in the office, or possibly even farther afield. You should know how to get to the phone closet by using the most direct route possible in the event of a telecommunications crisis. If you have a larger company, the phone system may even be sitting comfortably in a corner of the room that houses the company's computer servers.

If nobody has any clue where your phone system is stashed, you need to follow the cabling from your phone to the point where it meets up with all the other phone lines and becomes an indistinguishable mass of wires. That mass ties into a large panel or a large plastic or metal box with pretty green and yellow lights flashing on it. When you see this, you have probably arrived at your phone closet.

Phone rooms are frequently locked to prevent corporate espionage or someone from backing up into the wiring blocks with a wrench, taking out ten people's lines in the process. If your company does lock the phone closet, make sure you know where the key is so that you won't be scrambling when you need to get into it to check things out.

Putting a Name with a Face: Identifying Your System

Before you can do much, you need to determine the type of phone system you have. Fortunately for all of us, there are only a handful of types of phone systems. Each type is easy to identify by its features and requirements. The following sections give you the lowdown on the pros and cons of each type.

No phone system

The bottom end of the food chain is to not have a phone system at all. Small businesses that only have one or two lines, but no central room that pulls all the phone lines into a control box of any kind, have no phone system. If you follow your phone lines across the wall, eventually they pass through the wall to a gray or brown metal box outside the building, and then off to a telephone pole.

If you have no system, the phone on your desk, whether it's a single-line phone or a multiple-line phone, has the same phone number. If someone is on line one, everyone else can see it lit and knows not to push that button and start dialing. These phones are simple to use and you do not generally have to dial 9 for an outside line.

Pros

The greatest thing about not having a phone system is that if anyone ever has a problem dialing out, there is a 99 percent chance that the problem has nothing to do with you. This fact makes troubleshooting problems a breeze. Simply plug in a phone that you know works into a jack that you suspect is bad. If you still can't dial out, you now move to 99.999 percent certainty that the problem has nothing to do with your phones. Having no system involves the least hassle and lowest maintenance cost.

If you decide to upgrade at any time, you don't have to worry about the limitations of your existing system or the money you have already invested in it, because you don't have an existing system and have made no investment. It is financially and technically easier to start from scratch, installing a brand-new system, than having to deal with limitations of an outdated and costly system that can't be expanded any more.

Cons

If you lack a phone system, you either have to rely on an answering machine to take your messages or sign up for a voicemail service provided by your local carrier. If your business is more of a hobby, or has a Bohemian twist to it, that may be fine, but you need to have a voicemail service of some sort if you want to project a professional image. Along with the lack of voicemail, you may not have conference call or call-forwarding services (call forwarding can be great, especially after hours). Many local carries do offer packages that include voicemail, conference calling, and call-forwarding services.

Your decision

If you're reading this book, chances are you're ready to upgrade to a phone system. If you're still on the fence, ask yourself whether you plan to add two more employees in the next six months. Now think about whether you will add employees in the next five years. The next ten years? Do you need a professional phone image so that potential customers get the impression that your business is larger?

Centrex or Centranet

If you don't have a phone system and you don't want to pay to buy and maintain one, you do have an option. Most local carriers have a business service option. The Baby Bell companies like Bell Atlantic or Pacific Bell call the service *Centrex.* Verizon calls its service *Centranet.* Whatever it's called, the product is essentially the same.

The local carrier offers Centrex (or Centranet) lines that enable the features of a phone system without all the hassle of having one. You probably have to dial 9 for an outside line, you can have voicemail, conference calling up to two calls, and a whole list of other features that you may never need or want. If you are concerned about projecting a professional business image, but do not want to pay for a phone system, this option may be your best bet.

Pros

You don't have to buy any additional hardware and you can pick and choose the features you want and will use. These services also prevent you from having to buy and maintain hardware. Local carriers usually start businesses out with at least three phone lines, and reserve four or five more lines so that you can quickly add new numbers if you decide you need more. If your main line ends in -3450, and you have -3451, -3452 and -3453, you will probably have -3454, -3455, and -3456 waiting in the wings for you to activate when you need them. Because the numbers are already on reserve for you, it makes it easy to quickly add more lines in your office.

Cons

As your company grows, you will need more services. Eventually, you will want the hardware in your building so that you can repair and modify the system yourself. Paying your local carrier for the services every month will eventually be more expensive than buying a phone system and providing them yourself. Installing a phone system allows you to select as many features as you like and only pay for them once.

Your decision

If you don't believe that you will ever have more than 15 employees, and you really hate to deal with phone systems, stick with Centrex or Centranet service. If your business bounces up and you double in size, you can always buy a phone system without losing money on incompatible hardware. Eventually, you will need to bite the bullet and make the jump to a phone system because the cost of paying for individual phone lines and Centrex/Centranet features is higher than buying a system and getting a dedicated circuit.

The key system

If you have a *key system,* you have a phone closet with a metal or plastic box in it that all the phone lines in your office connect into. One feature that identifies a key system is that you don't need to dial 9 for an outside line. Some larger non-key systems dial the 9 for you with a feature called *assumed dial 9,* but the key system is the only one that doesn't require it at all.

Although key systems may have features like voicemail and conference calling, they're still fairly basic systems.

Old phone systems don't die; they just get resold

If you're buying your first phone system or replacing an existing phone system, your hardware vendor can help reduce costs. The technology won't be cutting edge, but then again, you may not need a system that can scale up to 672 phone lines. If you don't foresee that you will ever grow to need more than 48 lines, your hardware vendor may have a used or refurbished system that works just fine for you. Before you buy a used or refurbished system, you need to ensure the following:

- The hardware vendor can and will maintain the system.
- You can still easily purchase parts for the system.
- The expansion cards or the phones that connect into it (yes, you may have to buy new phones) don't cost so much that buying new phones is cheaper.

Along the same line, if you are completely replacing your phone system, call your phone vendor to see whether you can trade in your old system. For every business moving to a 200-station system, there is some little start-up in need of a system for 20 people. Your hardware vendor may be able to help you sell your old system, or may buy it from you for a fraction of what you paid for it (but still more than you would get if you threw it away). If you can buy a system for half the retail cost and sell it for a third of what you paid for it, that's smart business.

Key systems, as well as all telephones, phone systems, and wiring that reside at your office are referred to as *Customer Premise Equipment* or *CPE.* The term CPE may be thrown around when you are chatting with your local or long-distance carrier, so don't let it shake you. The term refers to the hardware at your building that you are responsible for. This is generally the last piece of hardware that is thoroughly worked through when troubleshooting an issue. When you hear the sentence, "Everything is clean here; we have isolated the issue to your CPE," that means you need to call out your hardware vendor to test your phone system.

If you have a key system, the dial tone you hear when you pick up a line is sent to you directly from your local carrier. If you ever have a problem with your phones (say you pick up the handset and you hear nothing but dead air), the issue is probably caused by your local carrier.

Pros

You own your own phone system and don't have to wait for anyone else to repair your issues on their timetable.

The main benefit of a key system is that it enables the telephone at your desk to access a larger pool of phone lines. If you have a three-line telephone *without* a key system that goes directly to your local carrier, you can only access the three lines plugged directly into that phone. If the lines in your stand-alone three-line phone are 206-XXX-1200, -1201, and -1202, for example, you can't receive a call on that phone from someone calling on line -1204.

A key system enables you to access all the phone lines for your company. You can program the key system to automatically route a received call to an operator phone or directly to a desk. Calls can also be transferred between phones connected to the key system, freeing up the initial phone that transferred it to receive another call. Even a small key system enables you to manage more lines and have increased functionality when compared to multiple-line phones used without a phone system. Owning your own key system prevents you from paying a monthly charge for Centrex or Centranet features.

Cons

Key systems are generally not very robust and may not have all the features you require. If you want to set up elaborate time-of-day routing features to automatically route calls to voicemail after hours, or if you want to have large-party conference calling, you may not get what you need from a key system. Because key systems are on the lower end of the telecom spectrum, you may not have a ton of room for expansion before you end up having to upgrade to a larger system.

Your decision

Take a look at your business and try to determine how many more phone lines you will need in the next 12 months, 5 years, and 10 years. When you have a rough idea of your needs, call your hardware vendor for a consultation. If your return on investment is more than six months, you may want to look for other options. If you suddenly hit an unexpected growth spurt and are forced into a larger system, you may have enough revenue coming in to buy that new phone system you *really* want.

The Post Branch Exchange: PBX

A *Post Branch Exchange* is more robust than a key system and can handle any quantity of phone lines. You can drop in 24 phone lines, or 672, or more. It really doesn't pay to use a PBX for fewer than 24 phone lines unless you love to buy high-tech gadgets with more features than you will use. The most distinctive feature of a PBX system is that you have to dial 9 for an outside line. PBX systems also have more functions available in them than the average key system.

You may be able to connect phone lines from different carriers into the system and have the PBX choose which carrier to send each call to depending on the area code and phone number being dialing. If you have a better international rate on your MCI line, for example, you can have calls that start with 011 sent over your MCI lines. If your cost for calls in state is lower through AT&T, you can program all the area codes in your state into the PBX system so that those calls are sent over your AT&T lines. This feature is actually called *least cost routing* (LCR) and can be built into some phone systems. Talk to your hardware vendor if you think your company can use LCR.

Pros

PBX systems have features that may be desirable for your office for one reason or another. Aside from the usual LCR features, routing features, voicemail and three-way calling, your PBX system may also be able to use account codes or call blocking features:

- **Account codes:** Have you ever wanted to know who in your office made a specific call? Have you ever wanted to track the calls you make on behalf of a specific client or project? Well, some PBX systems enable you to do it. The system prompts you for a code whenever you dial out. Generally, the code is a two- to four-digit number that the PBX stores in its memory so you can run a report later. If everyone in the Sales Department uses the 2211 code when they dial out, you can group all of their calls to see where they are focusing their efforts. If you are an attorney and bill your clients for all expenses on their case, you can easily assign a code for each client and tally up the total you need to tack onto the bill at the end of the month. This feature is quite popular, and if your PBX system doesn't supply it, your local or long-distance carriers probably offer similar services.

- **Call blocking:** With this feature, you can use the LCR feature to route outgoing calls to a dead line. For example, if someone dials a number in the 900 or 976 area codes, you can block the calls from being completed. Local carriers offer 900 and 976 blocking, as well as international blocking, but this may not be enough protection for you. Your company may also want to prevent people from dialing the 1010 codes advertised on TV for the dime-a-minute rate, or you may want to block outbound calling to countries like Canada or the Caribbean, which aren't covered in a standard international block offered by your local carrier.

International blocks placed on your service by either a local or long-distance carrier only cover calls that are preceded with 011. This standard international block works fine if you are trying to prevent calls being made to places like Europe, Asia, and South America. If you want to block all calls made to anyplace other than the 50 U.S. states, you need to install additional safeguards in your PBX system. Countries like Canada, Puerto Rico, Guam, Jamaica, Barbados and most of the Caribbean can be reached by just dialing 1 and the area code for the country.

Cons

PBX systems may be very costly. The specific telephones that are required to work with them also cost a pretty penny. Also, expansion can be very costly, and finding parts or qualified service technicians for older models isn't always a snap.

Another consideration is that if you have a PBX system, the dial tone you hear is generated by your phone system. If your building loses power and you have a PBX system, all the phone lines that feed into the PBX system go down. Unless you have a battery-backup system on your phone system, you can't accept calls or dial out until the power returns and your PBX system has the time to boot up.

The solution to this problem is to keep a cheap, single-line phone in the office and ensure that you have at least one phone line that doesn't go through your PBX. Your fax line doesn't need to access voicemail or transfer calls, so it makes a great candidate for doing double duty. Simply plug that single-line phone into your fax line and at least make calls outbound until the power returns.

Test-driving a hybrid

In telecom parlance, a hybrid isn't an awkward-looking car that runs on both gas and electricity. But it is a combination of two systems. Hybrid phone systems merge together the functionality of a key system and a PBX system. Companies that don't want to make the jump to a full PBX, because of either the cost or complexity, but want the luxury of dialing 9 before every call, can buy a hybrid and get the best of both worlds.

Your decision

If you're looking to buy a PBX system or want to upgrade your existing PBX unit, call a few hardware vendors for feedback. Your vendor may have a sweet spot in his heart for the make and model you own, but everyone else might see it as an expensive and brittle dinosaur. Before you go and throw down $15,000 or more on a PBX system and the accompanying phones, do some research and ensure that it matches your growth projections for the near and long term.

Multiplexers

A *multiplexer* is a device that receives a dedicated circuit from your carrier and breaks the circuit into individual lines that you can use to send to your desk phones.

In simpler terms, your carrier will charge you less to pull one fat cable into your building than it charges to pull 24 regular-sized phone lines. When the fat cable is in your office, you need a piece of hardware to break it down into 24 smaller lines that you can actually use. Multiplexers (also called *MUXes* or *channel banks*) are identified by the size of circuit they receive in from your carrier, and can vary from a single T-1 with 24 phone lines to a DS-3 with 672 phone lines. Some multiplexers handle circuits that contain thousands or tens of thousands of phone lines.

A multiplexer is a piece of hardware you may have in addition to your phone system. The MUX doesn't have any three-way calling features or voicemail; all it does is break down the dedicated circuit into phone lines. Any features that you need on your phones are still supplied by your PBX.

Pros

You can now use a dedicated circuit and save on the monthly charge to bring in all those individual phone lines. (See the section, "Making the final cost comparison," later in this chapter.)

Cons

The MUX is one of the first pieces of hardware into which your dedicated circuit connects. It is a potential point of failure that can take down all service riding through it if either the dedicated circuit or the MUX crashes. Without the MUX, you can't access any of the individual phone lines from your dedicated circuit, and all the phone service provided on the dedicated circuit is lost.

Your decision

Do you go dedicated or stay switched? To find out the answer, you have to crunch the numbers. Keep reading.

Getting the Least You Need to Know about Your Phone System

You should have enough information about your phone system so that you can perform basic troubleshooting functions. The more you know, the fewer the crisis situations you'll face. You don't want to have to call out your hardware vendor and pay the hourly rate of a Ferrari mechanic for some minor issue. Instead, invite the tech to visit once, and take very good notes in the following areas:

- What are the different devices in the phone closet and what do they do?
- How do you power up and power down each of the devices in the phone closet? What is the worst thing that can happen if you suddenly turn on or off the devices in the phone closet? Perhaps a sudden power outage returns all the settings to the factory default, which means that you have to reprogram everything!
- What does turning one piece of hardware off and back on do to the rest of the system? Is there a certain sequence for rebooting devices in the system?
- What do the lights on the equipment mean? You probably see green, yellow, red, and possibly blue lights on telecom equipment. A brief course in what red lights flashing on the channel bank mean will go a long way in shortening your troubleshooting.
- Is there a button on the *channel service unit* (CSU) or phone system that you can push to generate a *loop* on the system? You really don't even need to know what a loop is; just note the button you need to push to make it happen. I cover loops in great detail in Chapter 14.

Deciding whether to Get Dedicated or Stay Switched

The one decision in telecom that can impact your bottom line the quickest is the decision to replace some of your regular business lines from your local carrier with a single, large dedicated circuit from your long-distance carrier. This decision has such a profound impact on your phone bill because the per-minute rate for calls on a dedicated circuit may make up half the price you are currently paying for regular phone lines.

Understanding switched system costs

In order to make the comparison easy to understand, here's a look at the standard costs you can expect to pay to set up a *switched* (that is *regular*) phone system. For the purpose of comparison, I assume that you have 24 phone lines and already have a phone system that can receive calls from a dedicated circuit and are adding one piece of hardware (either a free-standing multiplexer or a multiplexer card). To add phone lines to an existing phone system that meet these criteria, you can expect to pay the following charges:

- **A monthly recurring charge of approximately $35 per phone line:** This is your standard monthly invoice that comes from your local carrier for the 24 phone lines. Any additional services on the line would, of course, cost extra. For now, I assume that all the enhanced features you need are going to be taken care of by your on-site phone system. Check your local carrier's Web site or talk to a friendly customer service rep to get exact pricing. The monthly recurring charge for the 24 phone lines is $840.
- **A per-line installation fee of approximately $100 (one-time fee):** Your local carrier is going to run wires to your office and drop them in your phone room, and that means you're going to pay a charge for that work. This fee could be waived, slightly less, or considerably more, depending on your local carrier and the type of phone lines that you buy. Your budget should be about $2,400 for installation of 24 lines.

  Verify the cost of installation with your local carrier before this work is done!
- **Your per-minute cost multiplied by the average number of minutes you talk every month:** A small business with good activity may use an average of 40,000 minutes per month, so I use that number as the baseline in this example. For the sake of simplicity, I also estimate that you are paying an average of five cents per minute for your calls. You should

check your current long-distance contract or just divide the total cost by total minutes on your invoice to get your true estimate of the blended cost per minute. The minutes and the cost per minute give you an expected average of $2,000.

- **PICC fee:** The *primary interexchange carrier charge,* or PICC fee, is a government tariff that is assessed to business and residential phone lines. The charge is applied on a per-line basis and can vary from $1.25 per line for residential lines to $4.75 per line for some business lines. Every local carrier has its own rate schedule, so I can't give you a fair sense of the standard rate. You have to research this fee on your phone bill and apply the correct amount when you calculate the cost. For the sake of this example and for ease of calculation, I am making the PICC fee a flat $2 per line per month; so if you have 24 lines, that's an extra $48 per month.

Aside from these charges, the regular taxes you pay remain a fixed percentage of your usage. You may incur some additional maintenance fees for the hardware you need to buy, but as long as you buy quality equipment, it should be negligible in the overall scheme of things.

Understanding dedicated system costs

The costs you incur to install one dedicated circuit (such as a T-1 line) to replace the 24 phone lines of a switched system consist of the following:

- **Monthly cost for the dedicated circuit:** In order to get those rock-bottom prices for long distance, you need to be directly connected to your long-distance carrier's network. Your long-distance carrier probably doesn't live next door to you, and the monthly cost of your dedicated circuit is based on how many miles of wiring needs to be used to get to you. The cost of the access varies from about $150 if you are a few miles away to possibly as much as $5,000 per month if you live far away from civilization. On the whole, a common access fee, commonly called a local loop fee, for the circuit is about $250 per month.

  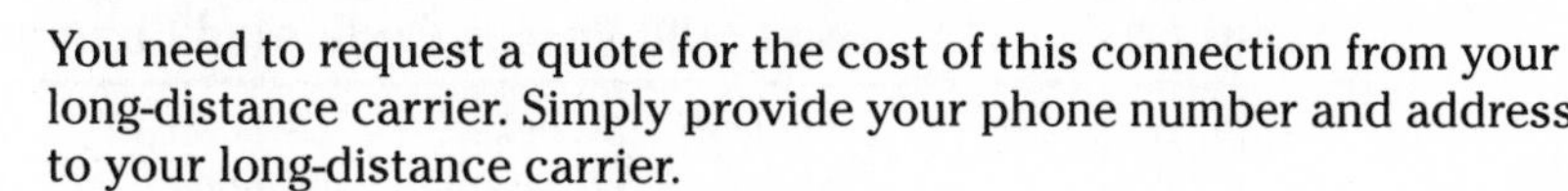
  You need to request a quote for the cost of this connection from your long-distance carrier. Simply provide your phone number and address to your long-distance carrier.

- **Installation cost for the dedicated circuit (one-time fee):** The copper wires that you rent from your long-distance carrier have to be designed, assembled, and manually connected. Sounds expensive, doesn't it? It is, and guess what! The cost is rolled up and passed on to your business in the form of an installation fee. The installation costs vary from carrier to carrier, but on average they run about $1,250 per dedicated circuit.

- **Additional hardware required (one-time fee):** The installation of a dedicated circuit requires you to purchase a multiplexer and a device called a *channel service unit* (CSU) that boosts the signal on your dedicated circuit and acts as a testing device. Chat with your hardware vendor to find a good deal on this hardware and to ensure that it's compatible with your existing phone system. If you're lucky, all you need to do is buy a special card for your phone system — everything else is automatically taken care of. Multiplexers range greatly in price from $5,000 to $6,000 for a Coastcom brand channel bank set up for voice calls, to a less expensive Carrier Access channel bank (you can find those on eBay for a Buy It Now price of $300). Whatever you buy, ensure that you have a hardware vendor who can install it and will service it. For this example, I assume that you got a good deal on a used multiplexer and had to pay only $1,500.
- **Installation and programming of new hardware:** When you buy the hardware, you still have to contract a hardware vendor to install it. As an average cost, I estimate that you'll spend $450 to install, test, and turn on the new equipment.

  Get a quote before you hire a company to do this work.
- **Your dedicated rate per minute:** I am giving a conservative estimate on this one by saying that you will be charged 2.5 cents per minute for outbound calls on your dedicated circuit with the exact same call volume of 40,000 minutes per month.

Making the final cost comparison

Table 2-1 includes all the charges I've mentioned in the previous section, listed as either monthly charges or installation fees. For example, in the first and second columns, I show what you will pay for switched service. In the third and fourth columns, I list the monthly charges and installation fees for a dedicated circuit.

Table 2-1 Switched Versus Dedicated Comparison

Item	*Switched Monthly Fees*	*Switched Installation Fees*	*Dedicated Monthly Fees*	*Dedicated Installation Fees*
Line charges	$840	$2,400	$250	$1,250
Hardware purchase	N/A	N/A	N/A	$1,500
Hardware installation	N/A	N/A	N/A	$450

Item	*Switched Monthly Fees*	*Switched Installation Fees*	*Dedicated Monthly Fees*	*Dedicated Installation Fees*
PICC line fees	$48	NA	NA	NA
Monthly usage costs (40,000 minutes)	$2,000	NA	$1,000	NA
Total Cost	$2,888	$2,400	$1,250	$3,200

After reviewing the two sides of the chart, you can see that adding a dedicated circuit costs $800 more to install than a switched system, but saves $1,638 every month in fees. In my example, the answer is simple; you can eventually recover the additional up-front cost in the very first month and still end up saving $838. The decision to move to a dedicated circuit is still a good idea even if you won't recover your upfront costs until six months after installation. If you won't reach a break-even point for a year or so, you may decide to stick with what you have.

Dedicated circuits are available in packages of 24 phone lines called a T-1 or DS-1, but it may be financially advantageous to order a T-1 line even if you use fewer than 24 switched lines. There is generally a break-even point at about 17 phone lines where it is better to buy a T-1 line than to stick with a switched system. Of course, every business scenario is different, so you should enter all the relevant information into Table 2-1 to see what works for your company.

Understanding and Preventing Fraud

The intent of telecom fraud is to gain access to a long-distance carrier's network by using your connection. After a fraudulent company is on the inside, it sells minutes to people at a discounted rate. The scenario is generally a *call-sell* operation. In this scenario, anyone on the street can pay $20 to call anywhere in the world for as long as they want. Because the person giving access to the call isn't paying for it, any money received is pure profit. There are many types of telecom fraud that can affect your business, but two varieties are more common and more costly than all the others.

Fraud is a concern in all aspects of life, but in the realm of telecom, a single episode can quickly cost you tens — or *hundreds of thousands* of dollars. No joke! The worst aspect of fraud is that even if nobody in your company made the illegal calls, you still have a legal obligation to pay for them. When you alert your carrier to the fact that your phone system or calling cards have been compromised, the carrier can help you identify and stop the current breach, but it has no obligation to credit you any money for the calls you claim are fraudulent. Any financial leniency you are given by a carrier is a professional courtesy unless your carrier has offered fraud protection within your contract for service. Always check your phone bill for international calls, because they are the largest targets for people committing fraud. Finding international calls to countries you can't find on a map at hours you are not even open for business is a huge red flag.

Stopping PBX hacks

As you read this, people all over the world are dialing toll-free numbers with their computers. When the toll-free number connects to a phone system, it begins hacking around to see whether it can access an outside line. If it can access an outside line, the computer then makes a test call to an international phone number. If the call completes fine, the series of codes and prompts used to access the outside line is recorded and sold on the streets by garden-variety thieves and organized criminals. This process of connecting into your phone system by way of a toll-free number and then dialing out is called *PBX hacking.*

The first thing to do to prevent your PBX system from being hacked is to eliminate your phone system's ability to *reoriginate dial tone.* That is, you want to prevent a scenario in which you receive an inbound toll-free call that can seize an outside line and receive dial tone. Obviously, accepting incoming calls is fine, but you don't want an inbound call to find its way to a dial tone.

If it's impossible to dial out of your system after a toll-free number accesses the system, you're in pretty good shape. Before the widespread use of cell-phones, re-originating dial tone from a phone system was a great feature used by sales staff while on the road. They could use a special company 800 number to call in for messages, and then press a code to dial out and return their calls — all without having to rack up expensive long-distance charges. But nowadays you have better options, such as using company-distributed mobile phones.

I strongly advise that you hunt down the toll-free dial tone reorigination option in your phone system and disable it. If you absolutely must keep it, at least have your phone system block international calls. The real money in telecom fraud is when you can sell minutes to locations that are expensive to call. Nobody is going to pay you $20 to talk to Mom in South Carolina, where the rate is usually about 7 cents a minute. The real market for telecom fraud is for people who want to call friends, relatives, and business associates in Afghanistan or the Kuala Lumpur, where the cost may be $2 per minute.

Cracking down on calling card fraud

Calling cards are being used less and less because mobile phones are now widely available at a low cost. However, calling cards are still a major source of fraud. The calling cards I am talking about are not the prepaid calling cards that you buy at the convenience store when you are traveling; I am speaking about the calling cards you receive from your local or long-distance carrier. Prepaid calling cards are set up to only allow a fixed amount of usage before they are disabled, so there isn't much fraud exposure with them. For example, if you use a prepaid card to make a call in a phone booth and you accidentally leave the card in the booth, you lose the amount of money left on the card — no more, no less.

The calling cards that you receive from your long-distance carrier, on the other hand, have a preset limit based on a daily balance, weekly balance, or monthly balance. If the credit limit is set at $100 per day, you have a potential $3,000 exposure per month on that card if the wrong person gets his or her hands onto it. Most people do not use their calling cards very often and you may not notice that the card has been compromised until two months later when your carrier generates an invoice and mails it to you.

Calling cards used for fraud are acquired in many ways. The most common way that calling card information is gathered is at airports with a little trick called *shoulder surfing.* Shoulder surfing is a technique whereby someone looks over your shoulder at a pay phone and makes note of the toll-free access number for your calling card platform and your PIN number. The person watching you may be in the next phone booth over or on an observation deck with a pair of binoculars, calling off the numbers to a partner who writes them down. Some criminals are more ingenious and place tape recorders under pay phones to record the sound of the digits you dial, and disreputable hotel staff may record the same information as you call from your room and it is processed through the hotel phone system. Regardless of how the information is gathered, your card information is then either sold on the street until your card's limit is hit or it is used by career criminals to cover their tracks and make phone calls to people they do not want to be linked to.

This fraudulent activity can add up to be quite a bit of money rather quickly. Here are some steps you can take to protect yourself from calling card fraud:

- Call your carrier and find out the credit limit and term on your calling card. Is it a set amount per day, week, or month?
- Have the limit set at $50 per month. Most people with calling cards make between $15 and $25 in calls per month, so this $50 shouldn't leave you stuck without any minutes when you need them most.

- Figure out how to quickly raise the credit limit on your calling card in the event of an emergency. The moment you need to raise the limit on your card will probably be a stressful one, so you should know your expectation level before you have to make the call. If you are at the hospital because your wife is having a baby, or someone was in an accident and you need to call people from the lobby phone, you will need the limit raised in a very short amount of time and as painlessly as possible.

Calls from your calling cards are charged a higher per-minute rate than calls from your home or office. In fact, the rates you're charged may be two or three times what you would pay if you call the same place from your office. Check your rates before you go on that long trip to avoid an unpleasant surprise when you get your next invoice.

Negotiating the Best Telecom Deal

The one constant in the world of telecom is that everything is negotiable. If you plan to spend $100,000 in usage every month, you can probably press to get installation fees, as well as small monthly recurring charges, waived. Thousands of carriers out there want your business, so check out a handful of them to see what kind of a bidding war you can start. Here are some tips to help you on your way to finding the right telecom fit for your company.

Getting the timing right

The landscape of telecom has been getting more and more competitive, with rates dropping, so don't get stuck in a lifetime commitment. Opt for a contract that lasts only 12 months. With that in mind, the best time to start looking at a new carrier is probably about eight months before your contract is up. That may seem like a lot of time to kill, but investigating a new carrier and determining whether it has the pricing, service, and support that you require can take a bit of time.

If you have dedicated circuits, it will take almost two months to install the replacement circuits on the new carrier. It will also take a month or two to work through the wording and negotiate the contracts. This is especially true if you are a busy company and cannot devote all the required personnel in your company to focus on this one task. The four months it takes to actually make the move to activate service is probably the same amount of time you will spend looking at new companies and weeding them down to the one that interests you.

Of course, these are ballpark time frames that can vary quite a bit. The point is to allow more (not less) time. If your company is growing and has little time to meet on issues like this, it may take eight months to get everyone together who has to approve the change to meet enough to make a decision. You could also be in a remote location where dedicated circuits may take three months to install. You will have to crunch the numbers and see whether the savings are worth the effort. If you are only looking at a $200 savings per month on a $5,000 bill you can probably renegotiate that with your existing carrier. If it is a $10,000 savings on a $120,000 monthly invoice, you will probably want to pursue the new carrier.

Dealing with least cost routing over several carriers

If your company is large and can use several dedicated circuits, you can maximize your savings by adding on more long-distance carriers. The wonderful thing about long-distance carriers is the fact that they are like snow flakes; no two are exactly the same.

This rule also applies to their pricing plans. Every carrier has areas in the U.S. and internationally where calls are inexpensive and other areas where they are not so competitive. If you currently have two or three dedicated circuits and a good phone system, you can build a *least cost routing* (LCR) table into it that identifies which carrier has the best rate. Then route those calls by using the appropriate carrier. There are two main things to bear in mind when you are going to activate an additional carrier to use in this kind of routing:

- **Make sure you will meet your commitment levels on the old carrier.** Most long-distance contracts for dedicated service give you a great per-minute rate, but they require a monthly minimum usage from you. After a ramp-up period of 30 or 60 days, you have to bill out $2,000 in usage per month to get your 2.5 cents per minute rate to Texas (or wherever). If you don't make enough calls, something bad might happen.

  Carriers don't make a habit of charging shortfalls, but shortfalls could invalidate your contract. The carrier is within its rights to assess the shortfall to your company or raise your rates. If you bring in a second carrier to save some money on your international calls as well as select areas in the U.S., you need to be sure that when that traffic is moved, you can still hit your commitment level with your primary carrier. Can you imagine telling your employees to make *more* calls?

- **Ensure that you set up the LCR table correctly.** Long-distance carriers have many different breakouts for pricing. Some may charge you a flat rate for all calls within your state, and another flat rate for all calls out of state. Regardless of where you dial, what time of day you dial, what day of the week you dial, and so on, the rate is always the same. Other carriers may break down the pricing per state, or base it upon smaller geographic regions within each state called a *Local Access Transport Area* (LATA), or even base it on the specific local carrier that receives your call. You want to make sure that you understand the level at which each carrier bases its rate so that you can build your LCR table accordingly.

For more information on how long-distance carriers determine the per-minute rate, check out Chapter 6.

Chapter 3

Getting Around the Telecom Neighborhood

In This Chapter

- Identifying the neighborhood
- Understanding various call types
- Making local calls
- Making toll calls
- Making calls within your state
- Making out-of-state calls
- Making calls in outlying areas and internationally

You use terms like *local call* and *long-distance call* all the time, but you may not know what differentiates a local call from a long-distance call. The federal government sets strict guidelines to stipulate when a local carrier can complete a call, and when the carrier must send it to a long-distance carrier. The need for these distinctions began a long time ago, when each state was broken down into small, geographic neighborhoods called *LATAs* (Local Access Transport Areas). Put simply, a LATA is the local region that makes up local telephone service area.

When you understand how LATAs function, the rest of it comes naturally. This chapter helps you understand the structure of the world of telecom so that you can make informed decisions on what you need from your carrier.

Identifying Your Telecom Neighborhood

Your telecom neighborhood is defined by the LATA, or *Local Access Transport Area,* from which you're calling. Although LATAs are regional designations, they were initially drawn up based on total population for the geographic area. This is the reason that states like Texas and California with higher population densities have 10 or 15 LATAs apiece, and the entire state of Wyoming makes up a single LATA.

Understanding the role of local carriers and long-distance carriers in LATAs

No phone call can legally cross a LATA border unless a long-distance carrier is handling it. Your local carrier, be it Ameritech, Bell South, or some other company, must transfer the call to a long-distance company like MCI, Sprint, or AT&T in order to complete the call. Figure 3-1 gives you a general idea of how the LATAs are set up in your state.

LATAs are *not* area codes and don't make up any part of a telephone number. The geographic area of LATAs is much larger than the area covered by area codes. In many instances, area codes exist in more than one LATA.

LATA numbering and borders

Every LATA has a specific three-digit number to make it identifiable. Say you're talking about calls that are failing in Oklahoma, for example. You could ask whether the problem appears to be happening more in LATA 536 than in LATA 538 or whether the issue is restricted to just LATA 538. LATA numbers are generally lower in the Northeast, and the numbering increases as you move west across the U.S. toward California. As populations explode in North America, new LATAs are constructed. You can easily find newer LATAs because they are all numbered in the 900s, like LATA 961 in Texas.

LATAs generally end at the state border, but there are many LATAs that cover more than one state. For example, half of LATA 956 is in Virginia and the other half is in Tennessee. The LATAs are not usually spit 50-50, but it is not that uncommon for a LATA to bleed over into a small section of a neighboring state.

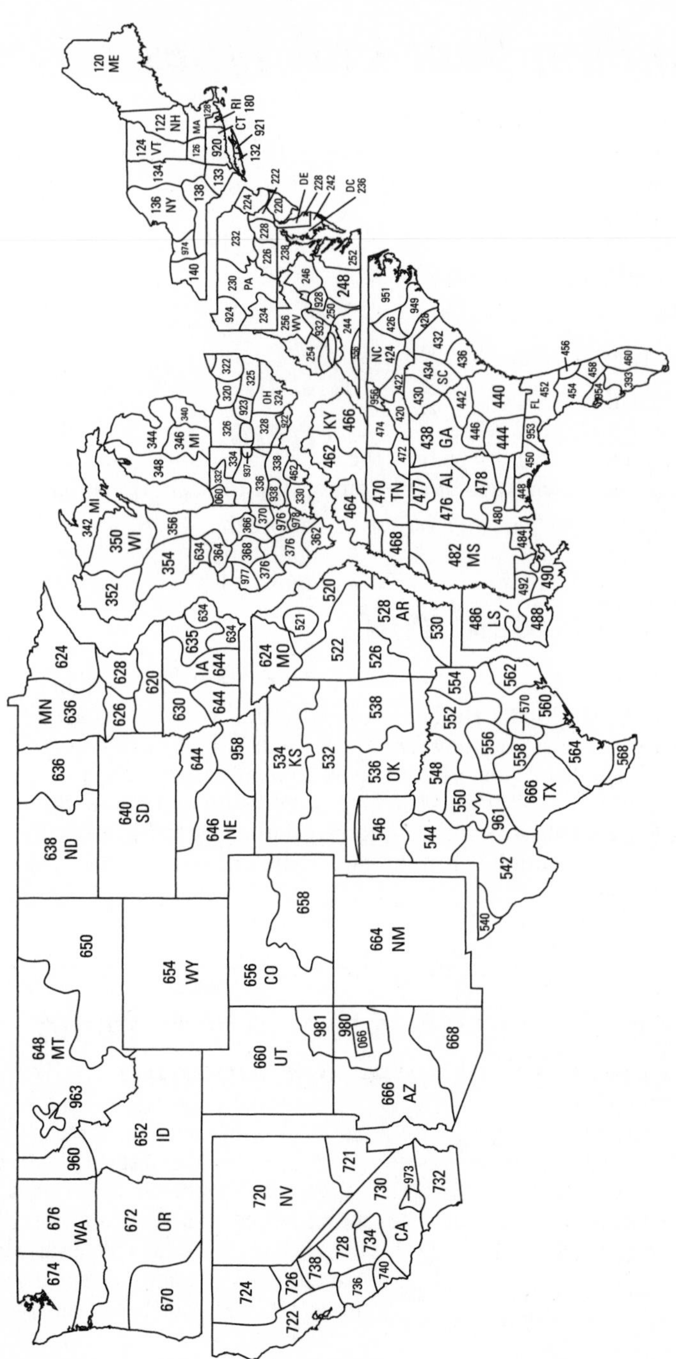

Figure 3-1: The network of LATA boundaries and LATA numbers for the U.S. is pretty complex.

Understanding Your Call Types

The phone calls you make fall into six categories:

- Local
- Local toll or intraLATA
- Intrastate
- Interstate
- Outlying area
- International

There is some uniformity to how the calls are grouped, but the individual carriers have their own ways to determine what they charge for calls within each category.

Making a local call

Any call that terminates less than 13 miles from where it originates is generally considered a local call. These are calls that your local carrier completes and, for residences, you may never see them individually listed on your phone bill. The front part of your telephone directory lists area codes and prefixes (your area code and the first three digits of a phone number) in your *local calling area.* If you call any phone number with the area code and first three digits of the phone number listed, you can rest assured you are dialing a local number.

Remembering what Stevie Mnemonic says about *inter* versus *intra*

Intrastate calls are local, toll, and long-distance calls that occur within the same state. Interstate calls are local, toll, and long-distance calls that occur between two states.

The prefix *intra* stands for within, and the prefix *inter* stands for between. If you have a difficult time remembering what is inter and what is intra, just think of the Internet. You can hop on the Internet and go to a Web site in South Africa, Japan, or anywhere else in the world, because the Internet crosses all borders. With that in mind, any call that crosses a border is an interstate or interLATA call, and any call that remains within the borders is an intrastate or intraLATA call.

Your local calling area is determined by your local carrier, and only covers an area within a 13-mile radius of you. This area doesn't include your entire LATA, which is how you get to make intraLATA calls.

Local calls for businesses are generally not free. The rate your business is being charged may be a penny per minute or less, but that little pittance adds up over time. Unless you're using residential lines for your business, expect to be charged for every call you make, even if you're just calling down the street to order pizza.

Introducing the LERG database

In case you are wondering how the local carrier can determine whether you're calling 13 miles or more away in about 100 milliseconds, I have a logical answer for you. The *Local Exchange Routing Guide,* or LERG, is a nationally updated database that receives information from and sends information to every local carrier in the U.S.

The LERG database takes the area code, along with the first three digits of every phone number and geographically links this data to the phone number's local carrier. (The combination of the area code and first three digits of the phone number is referred to as an *NPA-NXX.*)

The LERG database links this data to the physical location of the office that supplies service for that NPA-NXX. Whenever you make a phone call, your local carrier identifies the number you are dialing, checks its copy of the LERG database, and routes your call accordingly.

If you want a small look at a section of the LERG, see Table 3-1. As you can see in the table, every NPA-NXX is identified by city, state, and LATA. Each NPA-NXX is also identified by vertical and horizontal positions that act like GPS coordinates. The vertical and horizontal information is used by carriers to determine the distance between phone numbers. This is how your local carrier knows whether someone you are dialing is within the 13-mile radius for a local call, or whether you're placing an intraLATA call.

Table 3-1 LERG Database

NPA-NXX	*Vertical*	*Horizontal*	*City/State*	*LATA*
307-568	06668	06645	BASIN/WY	654
307-569	06668	06645	BASIN/WY	654
307-575	06981	05918	TORRINGTON/WY	646

(continued)

Table 3-1 *(continued)*

NPA-NXX	*Vertical*	*Horizontal*	*City/State*	*LATA*
307-576	06901	07067	LEIGHCNYN/WY	652
307-577	06918	06297	CASPER/WY	654
307-578	06674	06806	CODY/WY	654

Making local toll calls

Calls that you make to locations more than 13 miles away, but which are still within your LATA are called *local toll calls* or *intraLATA calls.* For years, in the late 1980s and early 1990s, these were some of the most expensive calls you could make. The local carriers were being forced to allow competition into their markets and they held on to this section of telephone traffic the longest.

Most local carriers rated local toll calls by increasing the cost of a call depending on the geographic distance between the origination and termination point, so a call that only traveled 10 miles was cheaper than one that crossed 200 miles. Okay, that makes sense. But that local toll call to a termination point 200 miles away cost *a whole lot* more than the local toll call to the next town over. How much is a whole lot more? Well, let me put it this way: You could call London, U.K., for less than it cost you to call to the farthest reaches of your LATA. In the mid-1990s, all local carriers finally allowed the option of having long-distance carriers automatically complete local toll calls. They generally have better rates than you would receive from your local carrier in this area, unless you are on some special promotional rate plan.

Making an intrastate call

An *intrastate call* terminates within the same state it originates, but it may not terminate within the same LATA. If you call from Dallas, Texas, to Houston, Texas, for example, you cross out of LATA 552 without calling outside the state of Texas.

The instant a call leaves a LATA, the call becomes a long-distance call — even if that call begins and ends within the same state. In the eyes and ears of the telecommunications industry, LATA borders are of greater significance than state borders.

A phone number by ANI other name . . .

In an industry that has an acronym for everything, you may not be surprised to discover that telecom community has even renamed your phone number. Your number is not just your number, or even your phone number, but an *ANI* (pronounced either *Annie* or *ay-enn-eye*).

ANI stands for *Automatic Number Identification,* and represents the entire ten digits of your phone number. You can use the term ANI rather than the phone number in any conversation with a representative at your carrier and the rep will automatically understand that you mean phone number.

If you want, the next time you chat with someone at the phone company, just throw in the term ANI so that the customer service agent knows you did your research and aren't someone to be trifled with.

Some long-distance carriers only have two general categories for pricing calls, *intrastate* and *interstate.* If this is the case with your carrier, your intraLATA, local toll, and intrastate calls are all charged at the same per-minute rate. The difference between intrastate long-distance costs from state to state is amazing. States like California may have an average intrastate rate of 5 cents per minute, but if you want to call from Miami, Florida, to Tallahassee, Florida, you can expect to pay about 15 cents per minute.

Making an interstate call

Any call that originates in one state and terminates in another is an *interstate call.* In general, the cost for interstate calling is less than the cost of intrastate calling.

If your main customer base is located in the same state your company is located, you may be better off financially if you relocate your office to another state! Making interstate calls can easily reduce your bill by up to 50 percent (and perhaps even more) of what you currently pay in intrastate charges.

Dealing with calls to outlying areas

Most long-distance companies in the U.S. have a gray area in their rates when it comes to interstate calls to Alaska and Hawaii. These states are frequently not included in domestic pricing, even though they are part of the domestic U.S. Along with Alaska and Hawaii, other U.S. principalities like the U.S. Virgin

Islands, Guam, and Puerto Rico are generally also put into this nebulous category, although they are technically international areas. Because none of these areas is directly attached to the lower 48 U.S. states, special carriers have to be contracted to transport the calls over the Pacific ocean, the Caribbean Sea, or through Canada. The markets in these areas are not large enough to bring in competition to drive the costs down, so the carriers that service these areas have to pay a little more overhead.

If you do business in any of the outlying areas, you should call your long-distance carrier and confirm your rates for calls both to and from these areas. Your outbound rate to call someone in Hilo, Hawaii, may not be that expensive, but the cost for someone there to call your toll-free number could cost your business twice as much. It is well worth making the call to prevent being hit with hundreds of dollars more in charges than you were expecting.

Making bona fide international calls

International calls are, technically, any call that terminates in a country other than the U.S. That being said, international calls come in two flavors, those that you have to start with 011, and those that you dial just like they were in the U.S.

Many countries can be reached by simply dialing **1+ area code + phone number**. Countries like Canada, Puerto Rico, Jamaica, St. Kitts, and many of the Caribbean nations are in the North American dialing plan. You can call individuals and businesses in these countries just as you call the rest of the U.S.

Of course, the rates are different than if you were calling domestically, but at least you don't need to dial 011.

Countries that are not in the North American dialing plan require you to dial: **011 + Country Code + City Code (if applicable) + phone number**. There is unfortunately no international standard for phone numbers. They could have a one-, two-, or three-digit country code, and possibly a one-, two-, or three-digit city code, before you get to the local phone number. Even the length of local phone numbers varies; they have six to eight digits. For example, call someone in the Seychelles Islands (located in the Indian Ocean, off the coast of Africa, in case you're not a geography whiz), you have to dial 011-248-XX-XXXX; to call London, U.K., you have to dial three more digits, even though the U.K. has a shorter country code and only a two-digit city code.

You can find a comprehensive list of all the country codes in the world at `www.countrycallingcodes.com` or look in the front of your local telephone directory.

The cost to call international destinations varies based on the following factors:

- **Which country you're calling:** The greater the price charged by the government or local carriers to terminate calls in a particular country, the more likely you will be to pay a king's ransom. If the foreign government decides it is a great source of revenue to pull money from these calls, you can expect excessive rates.
- **The geographic area within the country:** If you call a remote area in any country, you are going to pay more. As a basic rule, competition drives down the cost of international long-distance calls. For example, it's cheaper on some networks to call Manila, Philippines, than any other area in that country, because Manila, the capital city, is the most densely populated area of the country.
- **Whether you're calling a cellphone or *special service* number:** A special service number is a telephone number used for an entertainment purpose (see the sidebar, elsewhere in this chapter, "The secret's in the special service"). The reason these numbers cost a pretty penny is because companies built them specifically so they could inflate the cost as their source or revenue.

International price lists from long-distance carriers offer different rates based on these factors. For example, you may pay one rate to call the country in general, a different rate for calling a cellphone within the country, and still another rate if you call someone who lives in one of the major cities in the country.

Dialing internationally for the first time

Placing an international call based on what's printed on someone's business card isn't always the easiest thing in the world — especially if you don't do it every day. Companies print up business cards for domestic or regional use. If you travel to Europe for business and pick up someone's card, you may notice that European city codes look something like (071) or (045). You must dial 0 and then 71 before you dial the phone number. The *actual* city code may be 71 or 45, but in Europe the numeral 0 acts like 1 in the U.S., which you must enter before dialing an area code.

If you're calling a business associate in London and his or her business card says (071), you drop the 0 in front of the city code and dial 011 + 44 + 71 + *phone number.* But if you called that same individual from a phone in London, you'd dial 071 + *phone number.*

Blocking International Calls

If you contact your carrier to have a *standard international block* placed on your phone lines, you must remember that the block covers only outbound calls that require the user to dial **011**. International blocks don't block calls to Canada, the Caribbean, or any other locations in the North American dialing plan. If you see calls to Trinidad or St. Kitts, please do not get upset with your carrier; it's not your long-distance carrier's fault that international calls are still being completed by your employees. You should program your phone system to block the area codes you don't want contacted. If you do not have a phone system, you can always ask your carrier to assign account codes to your phone line so you can at least identify who in your office is making the calls. (See Chapter 1 if you are wondering about account codes.)

Playing the international dialing trivia game

The only two countries in the world with a one-digit country code are the United States (with country code 1), and Russia (with country code 7). If you call the U.S. from London, Paris, or Tokyo, you must dial the country's international dialing prefix (which, of course, changes by country), and then 1+area code+555-5555.

Most international city codes do not begin with a 0. The exception (there's always an exception) occurs in a few cities in Italy.

Mexico is the only country that's been broken up by long-distance carriers into seven different pricing bands. It is now more common to rate calls to Mexico by city, cellphone, and special service. Imagine dropping a stone into a pond; each ripple is farther from the center than the one inside it. This is how Mexico was initially set up, but in reverse. The geographic areas that are easiest to reach, like Tijuana, and the border towns by Texas are Band 1. As you go farther inland, the band numbers increase, along with the cost to complete the calls.

The most expensive international calls you can make are *ship-to-shore calls* and *shore-to-ship calls.* Both types of calls are equally pricey. These calls are also commonly called *Inmarsat (International Maritime Satellite Organization*) calls. The rate for these calls can top $25 per minute (no, I'm not kidding) to reach a ship in the Pacific. For that price, you would expect at least a rundown of your horoscope for the week or some lucky numbers you can choose for the lotto! Or you could save your money and just take that cruise instead!

Part II

Reviewing Telecom Products and Prices

In this part . . .

Part II gives you a solid introduction to all of the complex aspects of telecom. Chapter 4 starts by giving you the lowdown on dedicated phone service, introducing the lingo, the hardware requirements, and the pitfalls. I use Chapter 5 to introduce you to all of the complexities of toll-free numbers and provide solutions to the challenges you may face. I round out Part II with a rousing chapter about your phone bill, cluing you in on several billing options you can try, and giving you info about the standard, garden-variety items. This chapter is a great resource if you think you are being incorrectly charged for services.

Chapter 4

Understanding Dedicated Service Requirements

In This Chapter

- Understanding dedicated jargon
- Identifying what you need to supply for dedicated service
- Introducing channel banks and channel service units
- Knowing your carrier's responsibilities for dedicated service
- Getting the inside scoop on wiring

Dedicated service is telecom service provided through a dedicated circuit that only you have access to, spanning from your physical location to your carrier's network. It used to be a relatively rare telecom option, but it's becoming more common as hardware prices continue to drop and per-minute costs decline. If you're thinking about making the change to dedicated service, you should do a little research. Dedicated service requires you to take on a new series of responsibilities. For one thing, you'll be dealing with a whole new set of hardware. That means (ka-ching) buying the right equipment and making sure that you maintain it, as well.

This chapter goes over the details so that you know what's required of your company, as well as the carrier, when you order up a new dedicated circuit.

Understanding the Language of Dedicated Service

Before you can have a conversation about dedicated service, you have to understand the language. When you have the lingo down, the rest all falls into place. Here's the lowdown on the basic terminology:

- **Local loop:** The copper wire or fiberoptic cable that spans from your carrier to your office. Your carrier charges you a rental fee for the local loop every month and a one-time fee when it's installed.
- **Central Office or CO:** The building where your local carrier houses a large piece of hardware called a *Class 5 (local services) switch,* which routes calls, provides dial tone, and all the other telecom services you know and love. Also in this building are several tons of batteries to sustain the Class 5 switch in the case of a large-scale power outage.
- **Point of Presence or POP:** The building where your long-distance carrier houses a large piece of hardware called a *Class 4 (long-distance) switch,* which routes calls and provides all the long-distance services you know and love.
- **T-1 or DS-1:** A dedicated circuit that can be broken down into 24 individual channels, each of which can support a call. The T-1 circuit is also called a DS-1, and is the basis of all dedicated circuits; all voice circuits are broken down to T-1 level.
- **DS-3:** A dedicated circuit that holds 28 T-1s. This is a *big* cable. In fact, it doesn't even resemble a standard phone line that you plug into your phone; it looks more like the big, fat, round cable that connects into your television.
- **OC-3:** A dedicated circuit that holds three DS-3s. This is an *even bigger* big cable. It's a fiberoptic circuit and requires an *optical multiplexer* (or *optical MUX*) to receive it. An optical multiplexer is a device that puts the electrical DS-3 circuits into a fiberoptic cable. Optical multiplexers use light rather than electricity for signaling.
- **DS-0:** The atomic unit of telecommunications. A DS-0 represents the single channel necessary to carry one phone conversation. You need 24 DS-0 circuits to make up a T-1 or DS-1. That makes DS-0 the primary unit of measure for dedicated circuits in the United States.

- **E-1:** Outside the U.S., the basic unit for dedicated services. E-1 circuits carry 32 DS-0 circuits, 30 call channels, plus one channel for timing and one channel for signaling. The European standard was developed after the American DS-1 standard was set up.

Understanding Your Responsibilities When You Get a Dedicated Circuit

Because dedicated circuits are more complex than switched phone lines, they require more hardware to construct them. You don't just drop a dedicated phone circuit into your business without a *bit* of logistical planning. Your carrier only terminates circuits in accordance with industry standards — you can't get a partial T-1 circuit or half a DS-3 circuit.

The problem is that you can't make a phone call with a full T-1 dedicated circuit because the telephony network in the U.S. is set up for calls to be either 56KB (kilobytes) or 64KB. Trying to make a phone call with all 1.544 Mbps (megabits per second) of a T-1 circuit is like trying to take a drink of water out of a fire hose. You need to break down the flow of bits into smaller chunks so that your phone system can handle them.

The tool you use to break the large pipe into smaller pieces is a *multiplexer,* which connects into your existing phone system.

Getting all the equipment you need to make your dedicated circuit go

Here's a list of the equipment you need to make and receive calls on a dedicated circuit:

- **Phones:** You may already have these in place, or you may not need that many of them. Business models that provide calling card platforms or telemarketing services via an automated dialer handle most of the dialing with automated computer programs. These businesses may have three employees in the office, but the phone system may process thousands of calls an hour from customers calling in with calling cards or accessing prerecorded messages.
- **Breakout board or punch-down block:** These are plastic trays or racks that allow you to connect individual phone lines from your MUX or PBX to individual phones. Wiring and rewiring directly into a phone system presents the danger of damaging the connectors. These external racks allow changes to be made when one of your employees moves to a new desk or when a new fax line is installed, without having to continually press new wires into the PBX. Instead of destroying the port on an expensive phone system, it is better to short out a section on a cheap plastic breakout board.

- **Cabling to connect your phones to your phone system:** Unless you have some magic wireless setup in your office, you need to have someone physically run the cables from your phones back to the phone closet. From there, the cables connect either to a punch-down block (see previous bullet) or tap directly into the phone system. If you're adding a dedicated circuit to your existing system, you already have the wiring out to your phones in place.
- **A phone system — preferably a PBX system:** Smaller phone systems may not be able to handle dedicated circuits or may quickly hit their limit after they receive one dedicated circuit. A *post branch exchange* (PBX) system picks up where these smaller phone systems leave off. If you don't have a phone system yet, I recommend that you get a PBX system.

  If you already have a phone system but have never added a dedicated circuit to it, call your hardware vendor to ensure that your system can even handle the upgrade without additional hardware. Your phone system may have limitations or requirements that require a multiplexer. You simply need to know this information before you buy one.

- **A MUX or channel bank:** The MUX is a piece of hardware that receives the circuit from the carrier and parses it out into a smaller, usable size. Most MUXes break the circuit down one level, but some powerful MUXes can break the data all the way down to individual channels. For example, a T-1 MUX breaks the T-1 circuit down into 24 DS-0 circuits that you can use to make phone calls.

  A DS-3 MUX generally breaks a circuit down from DS-3 level to 28 T-1 circuits but goes no farther. You need an additional piece of hardware to break the T-1s down to DS-0 circuits so that you can actually use the circuit to make calls. This may seem like a pain, but if you have a DS-3 circuit, you may want some of your T-1 circuits to be used for an Internet circuit. With the two different MUXes, you can mix voice and data over your same dedicated circuit to your carrier. Several T-1 circuits could provide your office with Internet services and the others could go through a T-1 MUX to provide DS-0 circuits to your phone system for voice service.

- **Cabling from your phone system to your multiplexer:** This cabling may take the form of a single cable that connects from the phone system to the multiplexer (MUX), or it may consist of 24 sets of wires running from the MUX to a white rack called a breakout board or punch-down block. The breakout board is then wired into your PBX system or key system as individual phone lines.

- **Channel service unit (CSU):** The CSU acts as a testing point for any dedicated circuit and can be a stand-alone unit or a card that is integrated into your MUX.

  Although you can have a circuit without a CSU, I don't recommend it. The CSU is very helpful if your local carrier drops off your circuit on one end of your building and your phone system is five floors up and across the building (a likely scenario). The signal strength on a circuit does degrade the farther it has to be pushed, and the CSU's job is to boost the strength of the signal. When you buy a MUX, make sure it has a CSU; otherwise, you need to buy one to sit in front of the MUX. Adtran, Inc. (`www.adtran.com`) and Kentrox (`www.kentrox.com`) are manufacturers that make very nice CSUs. Remember that any time you buy a new piece of hardware, you also need cabling on the front and back end to connect it to your system.

- **Inside wiring:** The final thing you are responsible for is the wiring from your last piece of hardware, whether it's the CSU or the MUX, to the last piece of hardware from your carrier. This section of wiring, generally from the *NIU (network interface unit)* to your CSU, is commonly referred to as *inside wiring,* and it can be a point of contention when you negotiate with your carrier. Who is responsible for installing and maintaining inside wiring varies depending on how you set up your circuit. Your carrier may offer to provide the inside wiring or it may refuse to do so. Your responsibility depends entirely upon what the carrier is willing to do for you.

- **Network interface unit (NIU):** The network interface unit (NIU) is a small device that has a few pretty lights on it. It boosts the signal of a circuit and acts as a device for testing. The lights on the NIU tell you whether the circuit is in service, in alarm, or being tested. You should know where your NIU is and what it looks like because a time could come when your circuit is down when you may need to validate what's happening on the NIU. Chapter 13 discusses troubleshooting dedicated circuits.

  The NIU is the last piece of hardware that your carrier can easily test from a remote location. When a circuit reaches the NIU from the CSU, it continues on without much assistance until it terminates on the other side in your carrier's switch.

To help put all of this in context, Figure 4-1 shows how everything is connected.

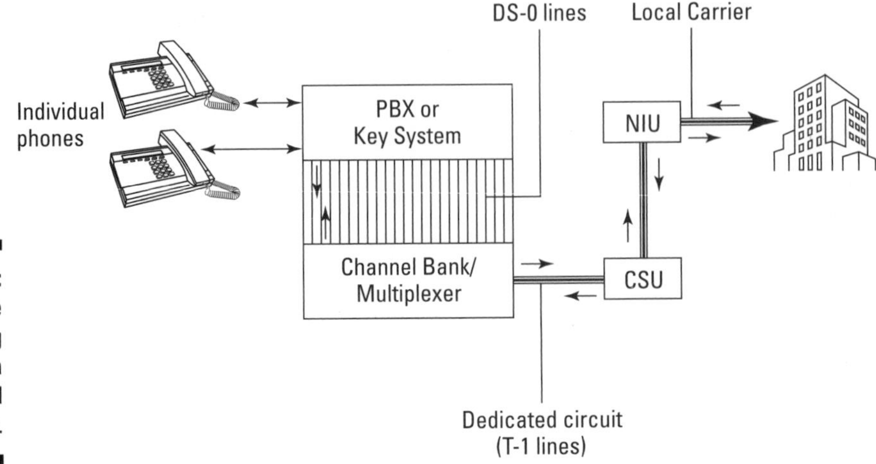

Figure 4-1: Hardware and cabling setup for a dedicated circuit.

Handling building-specific issues

If your business is housed in a new construction or in a very old building, your carrier may require you to provide logistical access and support. Be warned, trenching and plywood may be required:

- **If your building is brand new:** Your carrier may ask you to install a *backer board* in the main phone room. It sounds technical and complex, but the backer board is generally just a large piece of plywood attached to the wall that allows the carrier to secure its network interface unit (NIU) and any other hardware.
- **If your building is older and has many tenants with large telecom needs:** The conduit that brings all the wires and cables for phone service into your building may be full. You may have to install a new conduit so that your carrier can bring the circuit into your building. Although adding a new circuit doesn't sound that bad on the surface, it may be costly and time consuming, because your carrier may also require you to have the conduit buried from the edge of your building to the edge of your property. Although you won't personally have to dig the trench that goes from your building to wherever your carrier specifies (if you do dig that trench, you deserve a major raise), you will have to pay someone else to do it. In the best case, you have to dig a path through 10 feet of loose soil; in the worst case, you have to cut a path through 30 feet of asphalt in your parking lot.

Making sure your new CSU can be looped before you buy

The testing function of a CSU is solely dependent on it being *loopable.* This means that an electric current can be received by the CSU and sent back to the origination point. If your CSU is not loopable, you need to know that before there's a problem so that you can work around it.

If you ever have a problem with your dedicated circuit, one of the first things your carrier tries is *looping your CSU.* If the loop fails, the technician is likely to think he's isolated the problem to a defective CSU, when really the CSU is not capable of being looped. How embarrassing! Read the fine print before you buy a CSU.

Taking Responsibility for the Inside Wiring

In the previous section, "Getting all the equipment you need to make your dedicated circuit go," I make a passing mention of the inside wiring controversy. This section gives you the ins and outs of dealing with your carrier over this contentious issue. In the simplest terms, *inside wiring* is the wiring that goes between the network interface unit (NIU) to the channel service unit (CSU).

Knowing the facts about local carriers and inside wiring

To understand who is responsible when it comes to inside wiring, you need to understand the different methods of completing this connection. Then you need to accept that some things are out of your hands. The local carrier lays out what it believes its responsibilities are in terms of setting up the circuits, and although you might like to think that you can negotiate, the fact is that you may have to accept responsibilities (and the accompanying headaches) that you would really rather not deal with. The controversy over inside wiring stems from two important facts:

- **Fact 1:** The two most important physical locations in your building (from a telecom perspective, that is) aren't in the same place. You see, there's something called the *MPOE,* which is the carrier's initial entrance to your building. The MPOE is almost never in the same place in the building where you keep your equipment.

Demarc (pronounced *dee-mark*) stands for *demarcation point,* and the MPOE (pronounced *em-pow*) is short for *minimum point of entry.* There is usually a piece of hardware at the demarc that is installed by your carrier and represents the end of the carrier's domain. Your ultimate goal is to get your carrier to place the demarc in your phone room only a few feet away from your phone system. By wiring to your phone room, your carrier is providing you with an *extended demarc,* extending the carrier's responsibility only feet away from your hardware. The cable you plug into the demarc is your responsibility, along with everything else in your phone closet and all the wiring that runs into your office. The dictates of the local loop provider identify where it is allowed to drop the demarc. It may be at the MPOE, in your phone room, or anywhere in between.

The MPOE is usually a phone room or utility closet where your carrier drops a circuit if you do not specifically tell the carrier to put the circuit somewhere else (like, say, the demarc point). The MPOE is a room where all phone service enters your building. It is frequently locked, with only your local carrier or building management having the key. Representatives of local carriers commonly say that they will only drop a circuit at the MPOE, and that you are responsible for doing the rest of the connecting. You want to avoid hearing these words if you can (see Fact 2).

- **Fact 2:** The carrier is responsible for everything it installs. If your carrier agrees to go past the MPOE, then it is responsible for maintaining all the work it does past that point. That's better for you.

To give you an understanding of the significance of these facts, here's a scenario to consider:

Your carrier wires a circuit into the basement of your building and your hardware vendor takes care of the wiring in your office on the second floor, running cable from the phone closet in Suite 1 to your equipment in Suite 5. Sounds great, right? Well, not really. You still don't have a complete circuit.

You can easily overlook all the bits and pieces of this job because it all falls into the category of *inside wiring.* The person who wired from Suite 1 to Suite 5 on the second floor may say to you, "I finished the inside wiring," and you may perceive that to mean that he connected into the circuit in the basement when in reality he didn't. There is a lot of gray area in inside wiring, so make sure you are thorough when following up on it. You need to know which specific locations have been wired by each contingent of this massive project.

Someone must tie into the cable in the basement and then pull the cable up to connect the other end into the phone closet for your floor in Suite 1. That someone is determined through a series of negotiations, planning, pleas, prayers, and luck. If your carrier doesn't provide the wiring, your only option is contracting your hardware vendor to complete the connection. This isn't undesirable; it simply makes the vendor, rather than your carrier, responsible for the wiring. Confirming with your carrier that all the wiring connections have been completed will set your mind at ease. Most of the time, wiring is completed to your phone room without any problems. The challenge arrives if your building management is difficult and isn't working with your carrier. In this case, be prepared for weeks of negotiating the install of your circuit.

Addressing your options honestly

You only have three flavors of inside wiring options to choose from, depending on what your carrier either can or will provide:

- **Worst-case scenario — carrier wires to the MPOE only:** This is where your carrier, for one reason or another, will only install your circuit in the MPOE. The technician installs the circuit on the NIU (network interface unit), labels it with your *circuit ID* that acts as an inventory number with your carrier, or some kind of identifier, and then promptly leaves. If this is the case, you have to contract with your hardware vendor to wire into the NIU and run cable to the room with your phone equipment. You need to coordinate this effort with the office management because the MPOE may be locked.

  After the cable is run into your phone room, your hardware vendor should terminate the circuit on a phone jack (connector) of some sort. The circuit could possibly be wired directly into your CSU, but a direct connection of this nature makes replacing the CSU and troubleshooting the circuit a tad more difficult. You will have a much easier time if you can simply unplug the cable instead of unscrewing every wire from a piece of hardware.

- **Could-be-much-worse-case scenario — carrier installs an extended demarc:** In this scenario, your carrier does not drop the circuit at the MPOE, but actually moves the NIU to the phone room on your floor of the building, possibly even into your suite next to your CSU. This scenario gives you a great deal of protection against inside wiring issues because your carrier is now responsible for everything that is installed to your phone room. The circuit is still dropped and tagged with your circuit IDs so that you can identify them.

  If you already have an extended demarc in your suite, you may see other circuits listed on the hardware from other carriers. If you have an extended demarc, your hardware vendor needs to provide a small piece of cable from the extended demarc to a phone jack. Make sure the jack is also listed with the circuit ID; it connects into the NIU. Then all you need is a cable to connect from the jack to your CSU and you are ready to test.

- **Best-case scenario — carrier provides inside wiring to RJ-45 jack (T-1) or coax (DS-3) cable:** This scenario rocks because it places the greatest responsibility for maintenance on the carrier. The only piece of cable that is under your care is the section that plugs into your CSU on your hardware side and the phone jack on the carrier side.

Getting tips and reminders from the pros

So many factors are involved in who does the installation of all the inside wiring, and how you're charged for labor, materials, and maintenance. Here are a couple of reminders from the insiders:

- **Maybe it's out of your hands anyway.** Some buildings have very strict rules on wiring and have set up contracts with a company that handles all inside wiring issues. Other buildings are a bit more flexible and allow other companies to do the work as long as they have the proper clearance and are sufficiently bonded in case something goes horribly wrong. If this is the case with your building, your local carrier is probably already authorized to pull the cable to your phone room. If by some fluke they aren't, your building management will gladly refer you to a company that is authorized.
- **Consider all the costs.** Inside wiring is *not free* and can cost more money to install. If your carrier provides the wiring, you may have to pay out twice the amount of what you would pay your hardware vendor, but then your carrier is responsible for the wiring. If your hardware vendor runs the wiring, you may pay less now, but you will definitely feel the pain later, if the vendor has to come out six months later to fix it.
- **Find out your carrier's inside wiring policy.** Some carriers charge you to install it, but only claim any responsibility for maintenance if a problem occurs within 24 hours of installation. Some carriers give you free maintenance on it for 30 days. One constant in the inside wiring game is that if your carrier orders the inside wiring, it is responsible to fix it if a problem develops. They may be charging for the repair, but it's still their responsibility to initiate trouble reports with the local loop provider and press through to resolution.

Shh . . . Check the bill. All carriers have very complicated billing systems. Sometimes that can work in your favor. A carrier may very well have difficulty tracking things like inside wiring charges. Although you shouldn't count on *not* getting charged, the bottom line is that the fee you were supposed to get hit with for installation or maintenance may never show up on your bill.

Chapter 5

Meeting Toll-Free Service, the Red-Headed Stepchild of Telecom

In This Chapter

- Understanding the pros and cons of toll-free service
- Finding the best toll-free service for your business type
- Knowing which toll-free options your carrier provides
- Protecting your phone bill with toll-free blocking
- Discovering dedicated toll-free features

Toll-free numbers are everywhere today: You see 1-800-FLOWERS advertised on television or 1-888-GON-REFI popping up on the Web to entice surfers to call for mortgage refinancing. Heck, you can't pick up a can of soda without reading, `COMMENTS OR QUESTIONS, CALL 1-800-555-XXXX.` These are just a few examples of how a toll-free number can be used to help you take orders or provide customer service. The basic concept behind toll-free service is that the business that owns the 800 number pays for the calls that it receives; the toll-free part applies to the caller, not to the company.

You may wonder how a company can afford to pay for so many incoming long-distance calls. Well, the widespread use of toll-free numbers shows that, despite the cost, companies believe these numbers add value to their business. In this chapter, I introduce you to the ins and outs of toll-free service, including the fees involved. I tell you how to go about discussing options with your carrier, explain how toll-free numbers interface with your other telecom services, and show you where you may run into potential toll-free pitfalls.

Taking a Peak at Toll-Free Service Basics

You know that toll-free numbers put the financial burden on the receiving end of a phone call, rather than on the calling end. Often referred to as *reverse billing,* toll-free numbers are convenient for customers and also enable businesses to build name (and number recognition). The following sections give you some general info about how toll-free numbers function.

Understanding the physical properties of a toll-free number

A toll-free number is not like your home or office phone number. The line for that number doesn't physically exist anywhere. If your business has a toll-free number and you dial out from a line that receives those calls, the person receiving your call sees your business's local phone number, not the toll-free number, in his or her Caller ID window. This happens because toll-free numbers are an inbound service riding over your existing phone line, and must have a local phone number or dedicated circuit to be routed over to allow them to work.

Toll-free numbers sometimes do appear in the Caller ID window, however. This doesn't mean that everything I just said is a lie, just that when a toll-free number appears in your Caller ID window you're experiencing a little telecom sleight of hand. Telemarketing companies are required by law to display a legitimate Caller ID phone number. To accomplish this for all of their customers across the U.S., telemarketers use their dialing software to insert a simulated Caller ID number. What you see could be a local phone number of the order or customer service department, or it could be the toll-free number used for accepting orders. The number you see in your Caller ID isn't legitimate, because it doesn't physically exist on a phone line, but that's just a technicality. The main thing is that you can use the number to call the telemarketer and buy products or to tell the business to stop calling you. The industry term for this simulated phone number is *pseudo ANI* (the ANI part stands for *Automatic Number Identification* — telecom-speak for a phone number). If your dedicated circuit uses the ISDN protocol, your hardware may be able to input the pseudo ANI on the calls for you. All other protocols require your carrier to assign the pseudo ANI to your circuit through their normal order process. Turn to Chapter 8 for more information regarding dedicated circuit protocols.

Understanding the ring-to factor

Toll-free numbers are linked to ring-to numbers. Although toll-free numbers aren't physically associated with particular phones connected with a phone jack into the wall, you do receive the toll-free calls with a traditional telephone. Incoming calls get sent to your office or your home when your carrier directs the calls to either a *ring-to number* or a *dedicated circuit.*

A ring-to number is the regular phone line that is tied to a phone jack on your wall. Your carrier has a database that tells it how to route calls so that they end up in your office. If you have a ring-to number configuration, you need to know the phone number your toll free rings to in case you have to troubleshoot it, or order more toll-free numbers.

Discovering toll-free numbers in their natural habitat: The SMS database

Because toll-free numbers aren't tied to copper wires in someone's home or business, you probably wonder where they do live. All the toll-free numbers in the U.S. live as data in a series of computers that make up a huge national network called the *Service Management System* (SMS) database. This national archive lists every toll-free number in the U.S. and directs every call made to a toll-free number to the correct network to complete the call. In addition to the basic routing information, the database contains very specific information about individual toll-free numbers.

Dedicated circuits use trunk groups to receive toll-free calls instead of ring-to numbers. Your carrier designs your dedicated toll-free number to route to a trunk group specifying channels on your circuit that are to receive the incoming calls. Trunk groups can be limited to a single channel on a circuit, or may include several T-1s. Chapter 8 covers this information in detail.

Understanding the role of a RespOrg

The term *RespOrg* is an industry buzzword that refers to several pieces of toll-free-related minutiae. For example, a company that controls a toll-free number is called a RespOrg. The name is actually a reduction of the words *responsible organization.*

In the strictest sense, the company that holds RespOrg of a toll-free number is the only entity that can change and manipulate the setup of the number in the national database. A RespOrg document, more accurately called a *RespOrg Letter of Authorization* or *RespOrg LOA* is the paperwork you need to fill out in order to transfer a toll-free number to a new carrier. If you're speaking about the *RespOrg department*, then you're talking about the overworked group of people who work for a carrier. The people who work for the RespOrg department of a carrier live in a room insulated from the outside world by 6-foot thick walls of rebar-reinforced concrete and have only one phone line tied to a fax machine. Their sole job is the migration, design, and activation of toll-free numbers. Wild times. If you're interested in the RespOrg department or the RespOrg LOA, please hop over to Chapter 9, where you can find out all the nuances of both. If you want even more information, you can explore the 800 Service Management System (SMS/800) Web site (`www.sms800.com`). This site is the heart of all toll-free service in the U.S. and Canada.

Accepting Financial Responsibility for Wrong Numbers

A toll-free number is not a miraculous filter that instinctively blocks incoming calls you would prefer not to take. Unfortunately, you have to pay for all incoming calls within the geographical region you specified when you set up the toll-free number, including the calls from people who mistakenly dial your toll-free number. Just as old homes bring unwanted mail, every toll-free number has a history that can affect how many mistaken calls you get. The longer a toll-free number has been active for another company, the more people will still have the number listed on the five-year-old letterhead or the ten-year-old business cards.

You can also get calls that are simply misdialed. The caller may have entered 800 rather than 888, or transposed some digits. Generally, these calls are only a handful per month and amount to a few pennies, hardly worth doing anything about them. However, the calls appear on your invoice and you are expected to pay for them.

Because the true 800 numbers have been around the longest, they have the most baggage in terms of potential carry-over calls from previous owners. If you order a toll-free number that begins with 800, you can expect calls from people trying to get life insurance quotes, make payments on their cars, or ask for technical help on the turbo engine they just bought for their new scooter. If you really have a problem with these calls, you should request an 877 or 866 number for your toll-free service. They are newer and generally have less baggage.

This isn't your daddy's 800 number

A long time ago (you know, back in your daddy's day), all toll-free numbers started with 800. Now times have changed. With cheap long-distance calling, as well as a huge population of fax machines and cellphones, everyone (including your daddy) can have his own 800 number. And it seems like everyone decided to get one. When the supply of 800 numbers ran out, a new 888 exchange was released for toll-free calls. People loved the 888 numbers so much that in a few years the 888 inventory was running thin as well. So 877 and 866 were released as the new toll-free numbers, with plans to open 855 in the future. These new number exchanges mess up the old nickname for toll-free numbers. You can't really call them *800* numbers because toll-free numbers may have any number of configurations. To avoid confusion and avert disaster in the world of telecom, the generally accepted practice is to refer to the whole bunch of numbers — 800, 888, 877, 866, and even the yet-to-be-released 855 — as *8XX* numbers, or as I do in this book, just *toll-free* numbers.

If a business accidentally lists your toll-free number, or one that people easily confuse for your number, in their latest marketing blitz, you could end up receiving thousands of misdialed calls per day. Every one of these calls takes time to answer and resolve, even if all your employees say is, "Sorry, wrong number." A month later, when you receive your invoice you will also remember that you have to pay for them. If you see the handwriting on the wall that the wrong-number calls are going to eat up your time or your budget, you have two choices:

- **Block your toll-free number from the geographic area the calls originate.** If your business is in Seattle, Washington, and all your calls are coming from people trying to win a sweepstakes from a local radio station in Cincinnati, Ohio, just block access to your toll-free number from Ohio. As long as you don't have any existing or potential customers in the area you are blocking, you have nothing to fear. And you can always reinstate access from Ohio callers when the buzz dies down.
- **Cancel the toll-free number and get a new one.** I know it sounds brutal, but you may have no choice. You can't easily filter the wanted from the unwanted calls. In the end, you will have to make a business decision whether to keep the toll-free number and live with the added time and expense, or make a clean break and get a new number.

Coming to Terms with the Toll-Free Life Cycle

Toll-free numbers can be an immense source of frustration if you do not know where they are in their life. Here are just a few of the scenarios that you could be dealing with in regard to toll-free telephone service:

- You reserve a number and fail to use it during the time it is in reserved status, thereby losing access to the number, and enabling it to fall back to the rest of the world. Keep reading.
- You want to use a toll-free number has been cancelled and released to the world the same day. This is *not good.* Very bad things can happen. The following sections are for you.
- You already have toll-free numbers and don't plan on canceling them, but are changing carriers. If you fit this scenario, you should jump over to Chapter 9 to find out about ordering toll-free numbers. You face specific pitfalls in that task that you need to know about now.

- If you're not moving your numbers, but they're just not working for one reason or another, Chapter 14 is your destination.
- Finally, if you are planning on ordering new toll-free numbers or canceling old numbers, you are in the right place.

Conception

In the beginning, thousands of toll-free numbers wander *spare* around the national SMS database. As long as they are spare, anyone can decide to secure them by placing them in a *reserved* status. Now the reserved status isn't forever; actually, it only lasts 45 days. If your carrier doesn't activate the number before the 45th day, the toll-free number is released back into the spare pool and it becomes free game for anyone else. The good news is that for the 45 days the number is in reserved status, no other carrier can touch it in any way. No one can attempt to migrate the number away either through the usual channels or by force. The number simply exists in limbo until you activate it or let it go.

If you are ordering new toll-free numbers, don't be in a hurry to publish them. The reality is that we live in an imperfect world, and sometimes people transpose digits, advertising your number instead of theirs. Or maybe you wrote down the wrong number? (Never!) A frazzled business owner may have forgotten that a number that he reserved 45 days ago (and is currently printing a million business cards to promote) has moved back to the spare pool and (oh, horror) been picked up by you. Or maybe it was you who thought you took that number out of reserve? (Never!) In either case, the toll-free numbers you believe in your heart are yours may not be there (or you may decide you don't want them). Many people believe that the second they receive a fax about their new number they have received a guarantee of the number's availability, but that is not true. A number isn't yours until you can receive calls from it. Never publish a toll-free number until you have activated it, made a test call to it, and ensured that it rings through to the correct number! I cannot stress this point enough.

Birth

After your carrier finds your favorite toll-free number, you have to activate it. This is usually an easy task for your carrier and if the toll-free number is going to ring into a typical, everyday phone line, the process should take no more than about 24 hours. During that time frame, your carrier simply updates the national SMS database so that every carrier in the U.S. knows how to route the call. Anyone in the order-entry or provisioning department at your carrier can send the order off through an automated system to the national SMS database.

When your carrier updates the national database, it also updates its network to route the call to your home, office, cellphone, or wherever else you requested that the calls be sent. When all the systems are updated, presto! You can receive calls.

Adulthood

After a toll-free number is active, it goes to work. During this time, you can move the number to another carrier, point it to another location, or make any variety of adds, changes, or moves on it. Toll-free numbers don't wear down or become brittle with age; they just keep on working as long as you do. Only when a company goes out of business or is significantly reorganized, nullifying the number's purpose in life, is a toll-free number deactivated.

Demise

There is a standard procedure in the industry for decommissioning a toll-free number, and the next person who picks up the toll-free number would like you to follow it. When you shut down a toll-free number give it a proper burial by blocking the number and then canceling it.

Your toll-free number can bypass the three-month hold time for a cancel order and be immediately released into the spare pool. If your carrier issues a cancel order on your toll-free numbers and does not block it for three months first, very bad things can happen:

- **Gone is gone!** If you send the order to your carrier to take down the wrong toll-free number and the number becomes born again immediately, you may have no logistical or legal way to get it back.

  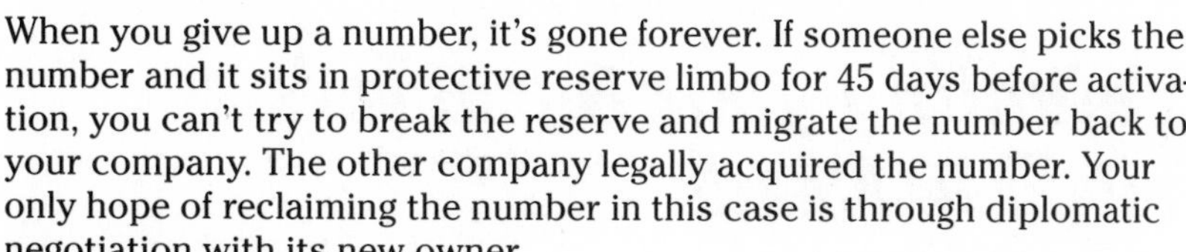

  When you give up a number, it's gone forever. If someone else picks the number and it sits in protective reserve limbo for 45 days before activation, you can't try to break the reserve and migrate the number back to your company. The other company legally acquired the number. Your only hope of reclaiming the number in this case is through diplomatic negotiation with its new owner.

- **Chaos ensues!** If you just want to be rid of a number and release it without following the protocol, you're potentially setting a stable world on a path to destruction. The system works because everyone plays by the same rules. When rules are broken, the whole telecommunications universe falls apart. Okay, I may be overstating things just a *tad,* but the point is that with toll-free numbers, patience is essential.

If, for any reason, you're done with a toll-free number, you must follow a pretty standard procedure. Give it a proper goodbye now so that it can be reborn later. Here's how the process should work:

1. You contact your carrier with an order to take down a toll-free number.
2. The carrier places a *block* on the number.

   Placing a block on a toll-free number simply prevents calls from completing to it.
3. Your carrier still has total control of the number for the 90 days the number stays in blocked status.

   Aside for ensuring that the block order shifts to a cancel order (see Step 4), for all intents and purposes, you're done.

   This is an important step, because if you sent the wrong number to be taken down by your carrier or you changed your mind for some other reason, you can still have it reactivated quickly and easily during that 90-day window.
4. After 90 days the block order automatically turns into a cancel order.
5. The toll-free number sits in a cancel status for another 90 days.
6. After these 90 days in limbo, it is reborn, becoming spare again. It's released for someone else to reserve and activate.

   The intention of this process is to give you three months to find your mistake if you cancelled the wrong number, as well as to allow a six-month cool-down before someone else can activate it. During that time, the people who have been calling the number for four years will realize that the number is out of service and will stop calling.

Following this professional courtesy prevents the next business that reserves the number from receiving wrong-number calls from people who wanted to speak to representatives of your business. Obviously, this is the protocol you hope the previous owner of your toll-free number followed, so be sure that you follow the golden rule.

Evaluating Your Business's Toll-Free Needs

Every business has its own, particular toll-free requirements, aspirations, and dreams. Your business must have certain features; other features would be nice, as long as they don't break the bank; and finally there are the features that you would love to have but may only exist in your dreams. The best way to ensure that you get everything you want is to list all of your preferences (and rank them accordingly) before you make the first call to your carrier. Or at least find out if what you want is available. The good news is that all carriers provide the same basic toll-free features. Not sure what you want, or even what's available? Read the following sections, which break down features based on business size and profile.

Small to midsize businesses

Business Profile: Companies in this category have 15 or fewer employees and fewer than 10 phone lines in the office. The main function of the business isn't telecom related and the level of service they provide is basic.

Recommendations: Regular, plain-vanilla toll-free service with no features may be all you need. You probably want your toll-free number to roll over to a group of lines set up by your local carrier so that when two people call in, the second person doesn't get a busy signal.

If you have a service business, you may need to make sure that the phones are answered after hours. You can get something called *Time-of-Day routing.* This service sends calls to voicemail after you close shop at night. If your existing phone system doesn't pick up calls after you close your shop, you can have your long-distance carrier send your toll-free calls to a different location, based on the time of day and day of week. What this means is that you can set up the toll-free number from 8:00 a.m. until 5:00 p.m., Monday through Friday to ring into your office, but at 5:01 p.m. you may want it to be pointed to an after-hours answering service, or to the cellphone of a designated on-call employee. You need to tell the carrier where to *repoint* calls from 5:01 p.m. to 7:59 a.m. every morning during the week and all day on the weekends. You also need to confirm your company's holiday schedule with your carrier so that you are all in agreement when your office will be closed. You may work a half-day on Christmas Eve, or the factory may shut down between Christmas and New Years: In either case, you need to cover all of your standard days off to make sure you are all on the same sheet of music. Of course, your phone system or local carrier may already cover this service, so make sure you know what features you already have before opting for more.

Small to midsize service businesses with several locations

Business Profile: Companies in this category provide a very high level of service within an industry that demands quick response times. These businesses have fewer than 20 employees per location. Cab companies, tow truck services, and plumbing/electrical repair companies are businesses that you might expect to find in this category.

Recommendations: *Geographic routing* (sometimes called just *geo routing*) is a great feature for small service-oriented businesses with multiple locations because it enables you to list a single toll-free number for all of your customers even though calls are routed to various office locations, depending on where your customer is calling from. For example, if you have a regional towing company with two locations in the Northwest and someone calls you from Sutherlin, Oregon, the call is routed to your office in Medford, Oregon. If someone else dials the exact same toll-free number from Seattle, Washington, the call is routed to your Tacoma, Washington, office. Geographic routing uses technology to project the same sense of continuity you get from a large business while also offering the personalized customer care you expect from a smaller company.

Depending on your business, you may set up geographic routing so that anyone who calls from a state in the eastern time zone rings to your Florida office. If someone calls the same toll-free number in the central time zone states, the call can be sent to your Colorado office, while calls in the pacific time zone ring to your Oregon location.

In addition to the geographic routing, you may also want Time-of-Day routing (see the previous section, "Small to midsize businesses") so that calls can roll over to the cellphones or pagers of your after-hours staff.

Telecom-centered businesses

Business Profile: Companies that fall into this category sell telecom products or services (calling cards, telemarketing services, and so on) as their primary or secondary product. If you find yourself ordering toll-free numbers in quantities of 100 at a time, or if your company's phone bill is 50 percent or more of your operating expense, your business falls into this group.

Typically, telecom-centered companies use a large quantity of toll-free numbers with basic features, although they may have a bunch of toll-free numbers with advanced features. The bottom line with these companies is that their toll-free numbers are their lifeblood, and without them they do not make any money. These companies generally have a complex phone switch that works 24 hours a day, 7 days a week, 365 days a year processing and tracking all calls. This consistency of service prevents any need for Time-of-Day or geographic routing, because telecom companies always want calls to come to one place — the switch. Your main concerns if you work in this type of business are pay phone surcharges and redundancy.

Understanding the origins of pay phone surcharges

Toll-free numbers are free for callers to use, but 20 years ago they were even more free. If you dialed a toll-free number from a pay phone, the call completed and the PSP didn't receive or expect any money for the call. PSPs didn't see a need to add the charge because cellphones were not common, and enough people were dropping 25 cents into pay phones that the PSPs were happy and rich. In the last 20 years, though, everything has changed and now fewer calls are made from pay phones. A larger percentage of those calls are made to toll-free numbers than to regular numbers. The PSPs realized this and lobbied Congress for a surcharge so they can stay in business. The fee started out as 23 cents, grew to 35 cents, and is now 55 cents per call.

If telecom is your business, you can understand why making your toll-free numbers work for you is essential. You can also understand why I can't give you all the information on this massive topic in one place. The following sections offer basic recommendations and tell you generally what you need to do to protect your business from catastrophic charges and network failures.

Recommendation 1: Factor in pay phone surcharges

Pay phone surcharges are federally mandated fees of 55 cents per call. The surcharges are assessed by pay phone owners (also called a *pay phone service providers* or *PSPs*) to the recipients of all toll-free calls made from pay phones.

Your company may need *pay phone blocking* if your hardware can't detect the information embedded in each call that identifies where it originated. If your business sells calling cards and you charge customers ten cents a minute to use one of your calling cards, that's one thing. But if a customer makes a two-minute call from a pay phone with a calling card, your business loses money — 35 cents, to be precise. That's not including the per-minute rate your carrier charges you for the call, or your overhead in printing, marketing, and customer service. In this situation, you have two possible solutions:

- **Get the infodigits.** You need to upgrade your phone system with a piece of hardware that can detect infodigits embedded in the call so that you can pass that 55-cent surcharge on to the caller. When choosing this route, remember to add some fine print on your calling cards alerting your customers to the additional fee from pay phones.
- **Block toll-free calls from pay phones.** You can tell your long-distance carrier to block access to your toll-free numbers from all pay phones. This option eliminates the surcharge problem entirely, but it also limits customer access to your service.

If most of your customers don't use pay phones, paying an additional fee to block toll-free pay phone calls may be overkill. On the other hand, if most of your customers do call from pay phones, not using this option could kill your company. Take a look at your customer base and determine what you need to do.

The pay phone indicator that is embedded in each call is referred to in the industry as the *ANI II* (pronounced *ann-EE-eye-eye*) or *ANI Infodigits.* ANI Infodigits are two-digit numbers that identify the origin of a call, be it a pay phone, jail, or operator assisted. Table 5-1 lists a few ANI IIs, as well as their descriptions. The ANI Infodigits are present in the overhead of every call, but only visible to you when using a dedicated circuit. Even then you will need the appropriate hardware with the correct protocol to identify and process the calls accordingly.

Table 5-1 ANI Infodigits

ANI II	*Description of Code*
00	Calls from plain old telephone service (POTS) lines are non-coin calls and require no special treatment.
06	Calls from hotels or motels served by PBX telephone systems. Receive detailed billing information, including the calling party's room number.
07	Special operator handling required.
23	Calls originated from a pay phone.
27	Calls originated from a pay phone where the pay phone provider does not supply an originating ANI.
29	Prison/Inmate Service. The ANI II digit pair 29 is used to designate lines within a confinement or detention facility that are intended for inmate/detainee use and require outward call screening and restriction. Federal prisons, state prisons, local prisons, juvenile facilities, immigration and naturalization detention facilities all fall into this category.

The information in Table 5-1 is only available to your phone system if you have the following:

- A dedicated circuit from a carrier's hardware that provides ANI Infodigits.
- A dedicated circuit using either ISDN or Feature Group D protocol.

Please check out Chapter 8 if you want more information about these protocols and how they work.

If you provide a service to the PSPs, such as operator service, you may be able to qualify for a *pay phone waiver* to exempt you from paying the surcharge. It doesn't make any sense for a PSP to contract with your business and then charge you 55 cents per call. You'll just turn around and charge that amount back to the PSP to cover your costs, and the employees in your accounts payable and accounts receivable departments will go on strike. Your carrier may make you sign a legal statement to protect itself from recriminations, but as long as your business has a contract with the PSPs authorizing that the fee is waived, you're in a good place.

Recommendation 2: Add redundancy

If toll-free numbers are the lifeblood of your business, you have them coming in on a dedicated circuit from your carrier so that you can maximize your profit (see Chapter 9), but what happens if your carrier has an outage and your circuit fails? If you don't have a solid disaster recovery plan, your business could be shut down for hours or days while network technicians repair the issue.

The best way to protect yourself against an outage is to have dedicated circuits from two different carriers, *and* have complete ownership (RespOrg) of your toll-free numbers. With this kind of power, you can redirect your toll-free numbers to your second carrier in a matter of minutes if lines from the first carrier fail. How you accomplish this is by becoming your own RespOrg.

Becoming your own RespOrg

If your business relies on toll-free numbers, you own hundreds of them, and you have dedicated circuits with more that one carrier, you should probably go one final step and make your company a RespOrg. Any company can become a recognized RespOrg and gain access to the national SMS database. This decision does require a bit of financial commitment, but it's worth it if toll-free numbers are a main part of your business.

To make your business a RespOrg, you must pay the $4,000 initial fee and pay for (and attend) mandatory training. After you pay the fees and complete the training, you get your very own RespOrg ID code, just like every other carrier, and you're off and running. You also have to pay a monthly fee for each toll-free number you have, as well as a fee whenever you touch a toll-free number to activate, cancel, migrate, or release it. If you're interested in making your business into a RespOrg, please check out the 800 Service Management System (`www.sms800.com`) for all the details, fees, and procedures.

If all this expense and training is a bit over your head, you can have almost all the benefits of a direct RespOrg relationship at a fraction of the cost by hiring a third-party RespOrg company. Third-party organizations have gone through the SMS training and are full-fledged RespOrgs that create a subaccount for your business. You still have your very own RespOrg ID code, but you do not have to go through any training because you don't have direct access to the national SMS database. Whenever you want to activate a toll-free number, simply contact the third-party RespOrg company and everything is taken care of for you. Using a third-party RespOrg is a great idea for people who make fewer than 2,500 additions, changes, or deletions per month to their toll-free numbers. If you're looking to go this route for your toll-free numbers, you can check out my favorite third-party RespOrg company, ATL Telecom Services (`www.atlc.com`).

Here are some of the benefits of becoming your own RespOrg:

- **You have direct access into the national SMS database 24 hours a day, 7 days a week.** You don't have to wait for a change order to move through the order system in your carrier. Rather than a 1-to-7-day time frame to make changes to your toll-free numbers, you can make changes in about 15 minutes.
- **You can use multiple long-distance carriers for all of your toll-free traffic.** All long-distance carriers are limited to routing calls over their own individual networks. Qwest won't send calls over AT&T lines for you, and MCI won't send calls to Sprint for you. Billing would be impossible, and no carrier wants to be responsible for troubleshooting calls that fail over another carrier's network. If you want to exploit the benefits of more than one carrier, your business must do it without the carriers' help.

  As your own RespOrg, you can choose which areas in the U.S. you want to route over which carrier. If you have dedicated circuits with both Sprint and MCI, and Sprint has a better rate for calls from Texas, you can point all calls from Texas over your Sprint circuit. At the same time, you can have all the calls from California route over the MCI circuit if MCI offers the best rates.
- **You can use a secondary carrier for redundancy.** If lightning strikes the switch for your long-distance carrier and turns it into a molten block of silicon and steel, that carrier is essentially worthless for a period of days or weeks when it comes to toll-free coverage from the affected area. If you have two long-distance carriers and both of them have your numbers set up to route into you, this catastrophe is a 15-minute inconvenience. All you need to do is go into the SMS database and update it so that all your toll-free traffic is routed through your other carrier.

- **Your numbers cannot be held hostage.** When you move your toll-free numbers from one carrier to another, your old carrier can reject your request to release them if you're in the middle of a billing dispute or because you are under contract. The carrier may hold onto your toll-free numbers and even take them out of service to force you to pay your bill. It sounds a lot like extortion, but these stipulations are generally written into any contract you sign for service. If your business is also a RespOrg, you own your numbers and control the carrier you send the traffic to. If you have service with both MCI and Sprint, and MCI shuts down your numbers down in the MCI network for any reason, you simply update the SMS database and send all the traffic to Sprint. In 15 minutes or so your toll-free numbers are back up and MCI can't use your toll-free numbers as a pawn against your business.

Large businesses with heavy usage and multiple locations

Business Profile: Companies with over 100 employees that spend more than $5,000 per month on toll-free phone service have a heightened need for redundancy, service, and disaster recovery. These companies have at least one dedicated circuit, and possibly several per location. Each regional office may have multiple customer service and sales support departments at various locations working in tandem.

Recommendation 1: Divide your calls in a percentage allocation

A *percentage allocation* enables you to split up where toll-free calls are sent based on a set percentage. If you have two offices that can take incoming phone orders, but one is twice as large as the other, you might want to have your toll-free number routed to the larger office 67 percent of the time and to the smaller office 33 percent of the time. Why not use your carrier to help you maximize your workforce? This solution works for almost any scenario where you have duplicate departments across locations.

Recommendation 2: Get geographic routing

Any business with several locations, big or small, needs *geographic routing.* Geographic routing enables you to provide localized service to your customers. For larger, national business, this solution is even more useful. If you have to service customers in all time zones, instead of having one office that works from 8 a.m. EST to 6 p.m. PST, you can run shifts that make sense. Nobody has to wake up at 5 a.m. PST or stay at work till 9 p.m. EST.

Recommendation 3: Become your own RespOrg

The control over your toll-free numbers and protection against having them held hostage is reason enough for any large company to become its own RespOrg. See the section, earlier in this chapter, called "Understanding the role of a RespOrg," for more information.

Recommendation 4: Get DNIS

DNIS (pronounced *dee-niss*) stands for *Dialed Number Identification Service,* and it is a feature available only on toll-free and 900 (9XX) numbers that terminate to dedicated circuits. The feature is actually a numeric code from two to ten digits embedded in your toll-free calls. The code enables your business to maintain a large number of toll-free numbers ringing into a single dedicated circuit. If you have 100 or 1,000 toll-free numbers that all receive only a handful of calls a day, you can program them into your phone system running over one dedicated circuit with only 24 available phone lines. The DNIS digits enable you to program a database that can automatically direct any inbound call to the appropriate department. As long as you don't have more than 24 calls active on your toll-free numbers at any given time, nobody receives a busy signal.

Here's how DNIS works. Say you have five toll-free numbers that your office has set up for specific departments:

800-555-3333 for Accounting, with a DNIS of 3333

800-555-4444 for Sales, with a DNIS of 4444

800-555-7777 for Customer Service, with a DNIS of 7777

800-555-8888 for Investor Relations, with a DNIS of 8888

800-555-5555 for the fax machine, with a DNIS of 5555

When you order dedicated toll-free numbers from your carrier, you tell the carrier the exact DNIS number you need assigned to every toll-free number. Check out Chapter 9 for all the specifics on ordering dedicated toll-free numbers.

You can choose any quantity and series of numbers for the DNIS, but you may want to make things easy on yourself. Common practice is to use a four-digit DNIS, and to base the DNIS on the last four digits of the toll-free number, as in the five examples I list in this section. If you have over a thousand toll-free numbers, you're more likely to have several numbers with the same last four digits. You may want to move to a seven- or ten-digit DNIS, possibly using the entire toll-free number as the DNIS or simply creating a unique DNIS for every toll-free number, with the DNIS and phone numbers having no resemblance to each other. The main reason to use part of the toll-free number as the DNIS is to make things easy. Someone has to program all the toll-free numbers and their associated DNIS numbers into your phone system. You cut down on possible programming errors if your technician does not have to cross-reference every toll-free number with a unique DNIS number.

Putting DNIS to work

This is what happens to an incoming call on 800-555-7777 for your Customer Service department when it hits your phone system:

1. Your carrier receives a call for 800-555-7777 and sends it to your dedicated circuit.
2. Your phone system sees the incoming call and sends a tone to your carrier to identify the DNIS of the toll-free number.
3. Your carrier sends the DNIS digits of 7777 to your phone system.
4. Your phone system checks the programming it has for DNIS 7777 and routes the call to the extension for Customer Service.
5. Your Customer Service department receives the call and your customer is taken care of.

DNIS is only available on dedicated circuits that use either E&M Wink or ISDN protocol. Some older phone systems may not be able to handle DNIS and you will have to partition your dedicated circuits and send calls for specific departments over specific channels. In this scenario, you will know that calls that come in on Channels 1 through 14 are for Customer Service; calls that come in on Channel 15 are for the fax; Channels 16 through 23 are for Sales, and Channel 24 is for Investor Relations. This partitioning limits the number of toll-free calls you can automatically route to a maximum of one for every channel. If you want to set up individual route paths for 50 employees in Sales, and you only have 24 channels available on your circuit, you are S.O.L (simply out of luck).

Recommendation 5: Get Caller ID (ANI delivery)

If you have toll-free numbers coming in on dedicated circuits and want to track the effectiveness of your marketing, or to have customers' account information available when they call in (even before your rep has a chance to say hello), you need ANI delivery. In the simplest terms, *ANI delivery* is Caller ID for dedicated circuits. The phone number that originates a toll-free call isn't always sent, and you will have to order the ANI delivery if you want to receive it. When you have this information coming into your phone system, you can begin tracking incoming calls. If you have just released a marketing blitz in Charlotte, North Carolina, you can then track all incoming calls from that area for the next seven to ten days. If you see a huge spike in incoming calls, and an associated percentage of sales, you know your marketing plan was a good investment.

If your phone system is a bit more advanced, you may be able to take the ANI delivery information and link it up to customers' accounts (as long as they are calling from their home or business). Using ANI delivery in this way enables your customer-care representatives to simply validate account information on-screen instead of spending time tracking it down or unintentionally introducing error by incorrectly reentering account information.

Just as with DNIS, ANI delivery is only available if your phone system is using either E&M Wink or ISDN protocols. If this is a feature you can use, reconfiguring your hardware or buying new hardware may be a cost-effective option if your existing system doesn't support either of these protocols. Call your hardware vendor for a quote and a timeline on how long it takes to bring this type of system together; determine the benefit to your company, and make your decision.

Recommendation 6: Get ANI Infodigits

If you have dedicated circuits and your customers frequently use pay phones to call your toll-free numbers, tracking the ANI Infodigits of the calls may be very beneficial to your business. You could be spending thousands of dollars per month in fees that are not accounted for in your business model. In the best-case scenario, you pay $5 or $10 more per month out of your profit. In the worst-case scenario, you are being hit for more money in pay phone fees every month than you are making in profit. If you're interested in ANI Infodigits, check out the recommendations I offer earlier in this chapter for telecom businesses.

Recommendation 7: Handle overflow

If you have multiple locations that perform the same functions (order taking, customer service, and so on), you will want to have your toll-free system set up to send overflow calls from one office to the next. If one office is inundated with calls, the calls overflow to the next office, which picks up the slack. This is also a very helpful solution if your hardware fails at one location; the overflow system automatically routes the calls to your next office without missing a beat.

Every carrier has different features and limitations when it comes to overflow, so you need to shop around. No industry standard is in place regarding overflow, and every network is built differently. So what you can do with a toll-free overflow plan from a carrier like WilTel you can't do with an overflow plan from MCI. Some of the limitations of overflow you can expect to find are:

- **You can only overflow between dedicated circuits that are installed in a geographically confined region.** All of your circuits must connect together in your carrier's network in the same location. If you have a circuit in Chicago, Illinois, and another circuit in Rockford, Illinois, you may be able to overflow between them. However, your carrier may not enable you to overflow calls between Rockford, Illinois, and Richmond, Virginia.

- **Yes, but you can overflow lots.** A variation of this limitation enables you to overflow between as many locations as your heart desires. So although you're confined to overflow within a geographic area, you can have two offices overflowing, or seven, or anything in between, without a problem.
- **Or you can't overflow much at all, but you can do so everywhere.** In yet another variation on this theme, you can overflow from an office in Miami, Florida, to an office in Seattle, Washington, but after overflowing two times (three locations, total), you may not be able to route the traffic to any other dedicated circuits.

Be sure to collect all of your information before you base your business on a service that may not be available. Read the fine print!

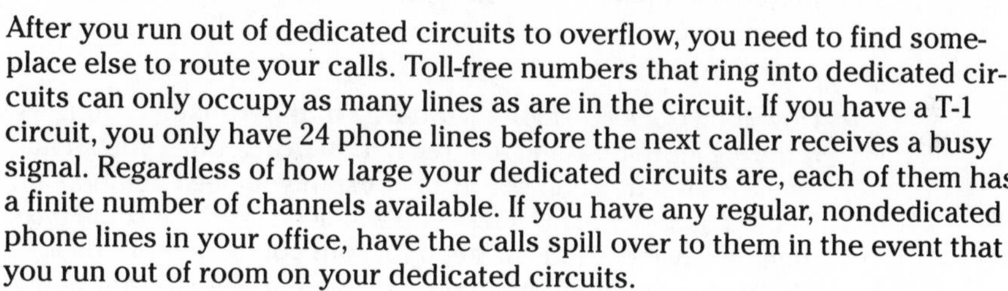

After you run out of dedicated circuits to overflow, you need to find someplace else to route your calls. Toll-free numbers that ring into dedicated circuits can only occupy as many lines as are in the circuit. If you have a T-1 circuit, you only have 24 phone lines before the next caller receives a busy signal. Regardless of how large your dedicated circuits are, each of them has a finite number of channels available. If you have any regular, nondedicated phone lines in your office, have the calls spill over to them in the event that you run out of room on your dedicated circuits.

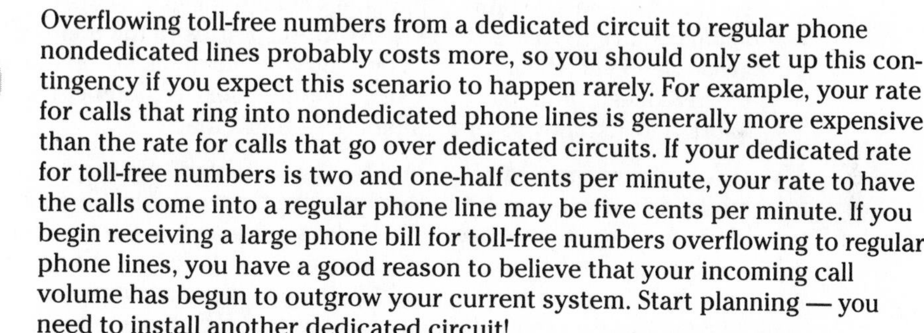

Overflowing toll-free numbers from a dedicated circuit to regular phone nondedicated lines probably costs more, so you should only set up this contingency if you expect this scenario to happen rarely. For example, your rate for calls that ring into nondedicated phone lines is generally more expensive than the rate for calls that go over dedicated circuits. If your dedicated rate for toll-free numbers is two and one-half cents per minute, your rate to have the calls come into a regular phone line may be five cents per minute. If you begin receiving a large phone bill for toll-free numbers overflowing to regular phone lines, you have a good reason to believe that your incoming call volume has begun to outgrow your current system. Start planning — you need to install another dedicated circuit!

Overflow for dedicated toll-free numbers is referred to by several names. If you want to sound like a pro, you can ask your carrier to have your calls *hunt* from one dedicated circuit to another. Everyone should understand what you are asking for. You also hear references to *dedicated trunk overflow*, or just *DTO*. Same thing.

Identifying Your Carrier's Available Toll-Free Services

When you've determined what your toll-free numbers must do to make your business work, call your carrier's sales rep and ask what's available. Your rep should have all the services suggested in this chapter, and probably a few more that the whizzes in Research and Development just created. When you have the features locked in, go deeper to determine what else your carrier can do for you. Some things you should ask about are:

- **Is redundancy for built in?** Your numbers may play a supporting role in your business, or they may *be* your business. If you have one toll-free number that receives four or five calls a day, redundancy may not be much of a concern. Ask your carrier how long it may take to send your calls to your cellphone or to another phone number if the phone lines currently receiving your toll-free calls fail. If you have more that one toll-free number, but fewer than 500, talk to your carrier about what kinds of disaster recovery and backup options are available. Redirecting your toll-free numbers to another office or to a cellphone may be your only option if your dedicated circuit dies, or your local carrier is experiencing an outage.
- **How long does reserving a toll-free number take?** As long as you're not ordering more than 500 toll-free numbers at a time, you should be able to reserve a toll-free number in 24 hours. If you need numbers quicker, project the quantity of numbers you will need for the next 30 days and order them in advance. You can keep them in your inventory by activating them with your carrier and pointing them to a phone line that nobody uses. If you need a toll-free number that spells something like `1-800-4MY-TAXI`, you may have to wait 24 to 48 hours for responses (if the number is even available). Everything depends on how your carrier is set up to handle these requests. You may just call a customer service number (toll free, of course) to instantly reserve numbers — that's that.
- **How quickly can a toll-free number be activated**? Again, your business model drives what length of time you deem acceptable. Companies that plan for product launches 30 to 45 days out may not worry about this as much as those that need fast turnaround times. If you have a pager company that assigns and activates toll-free numbers for every pager you sell, you can't wait a month. If you need your toll-free numbers up in hours rather than days, you need to confirm that your carrier can comply. Toll-free numbers assigned to regular, nondedicated phone lines generally take about 24 hours to activate, but setting up a dedicated circuit for toll-free calls may take 30 days. Know your time frame.
- **What is the carrier's response time if there is an outage?** It is never too early to meet the people who take care of your toll-free numbers when there is a problem. A day will come when your toll-free services don't perform to your standards. Your may have echo, static, or calls to your toll-free numbers may just fail to connect, leaving callers hearing a fast busy signal. Now is the best time to request the phone numbers for your support, as well as standard response times for your carrier. You can avoid frustration in the future if you know that they have a two-hour response time for a number that has 100 percent failure and an eight-hour response time for toll-free numbers with static or echo.

- **What online tools are available to customers?** Most carriers have secure Web sites that enable corporate customers to add, change, and block toll-free numbers. If your business is big enough, you can use the Web to get direct access. Using the Web can help you cut costs, manage your usage better, and even give you a competitive advantage on the service you can provide to your customers.

  Online tools are great and a lot of fun, but they also represent a considerable amount of investment from your carrier. If you have access to an online tool, you may be required to sign a large monthly commitment with your carrier, or the per-minute rate you pay may be a few pennies more. If the features are fun, but not integral to your business, you should weigh their benefits against the cost savings. I know that it is nice to see the number of calls and total minutes of those calls hitting each of your dedicated circuits, but so is staying in business, right? Such features can cost as much as $50,000 more per month.

- **What are the toll-free directory assistance options?** This is a very good question and most people do not choose to be listed in the National Toll-Free Directory Assistance database. Most people don't know how to access this directory, and if people don't know how to reach the service, why would you want to pay AT&T the $45 per month (or whatever AT&T is charging today) to be listed? Incidentally, if you want to access National Toll-Free Directory Assistance, the number is 800-555-1212.

Realizing the Cost of Enhanced Toll-Free Services

Toll-free numbers have the greatest variety of add-on services available, and every carrier assesses the fee in a different way. One carrier may charge a one-time installation fee when you activate a number, and another may activate numbers for free but collect a monthly recurring charge (MRC). Any feature you add onto your toll-free numbers may have a one-time nonrecurring charge (NRC) on it or a continual MRC, but is rarely free unless you negotiate to have the fees waived.

There are exceptions to every rule, but the two fees that you can always expect to see in relation to your toll-free numbers are:

- **The standard per-minute charge for usage.** Just like you pay a per-minute fee for your outbound calls, every toll-free call is charged for the time the call is active. The rate is generally a little more than what you pay for standard outbound calls. The only way you can avoid the per-minute fee for toll-free calls is if you have a package deal with a flat rate for all of your services. I can guarantee you that even if you have a package deal like that, somewhere in your contract or the contract of your carrier is a calculation for usage based on a per-minute cost.

Charges for toll-free calls vary! Your contract for toll-free service probably has a standard set rate plan for calls that only originate from the lower 48 states. If you are about to begin a campaign in Alaska, Hawaii, Canada, or any part of the Caribbean, check your rates before you send out the fliers. These areas may only cost 5 or 10 cents per minute to call out to, but inbound calls on your toll-free number may easily be 25 to 50 cents per minute. Ouch! Call your carrier to confirm the rates in writing before you receive a $5,000 phone bill for calls on your toll-free number from Puerto Rico.

- **Pay phone surcharges always apply.** Unless you have a contract with a pay phone provider, you will be required to pay the 55-cent pay phone surcharge on all calls to your toll-free number from pay phones.

Aside from these two fees that 99 percent of businesses find on their phone bill, there are other ways that carriers charge you for toll-free service. No two carriers are the same, so you will have to extract from them all the charges and fees that are in their ancillary fees list. They may also be buried in your contract, but it is worth the hour to read through all the sections. Some of the more common fees to ask about are:

- **Toll-free activation fee:** This may be a charge of a few dollars per toll-free number to build the routing plan and activate your number. If you have less than 20 toll-free numbers, you may be able to negotiate this fee to be waived. If you have more than 50 numbers, ask your carrier whether you can send the numbers over as a *bulk load* (you just list the phone numbers in an Excel spreadsheet or Word document instead of filling out the same information on multiple paper order forms) to reduce the charge. The bulk load process is probably automated and is definitely quicker.
- **Monthly recurring fee per toll-free number:** Even if you don't have any calls on a toll-free number you may be charged for simply having the number active. All the carriers are charged about 20 cents (the rate fluctuates) per number by the National SMS database for maintaining the records on each toll-free number.
- **Fees for enhanced routing features:** If you have Time-of-Day, or geographic routing on your toll-free numbers, you may be assessed a fee for that feature. There is, of course, no industry standard for this fee, but you should ask to see if there are any of the following fees on your enhanced toll-free features:
    - **One-time installation fee per toll-free number:** This may be as little as $1 per number or $10 per number. It is generally not a massive fee, but if you have thousands of numbers, the cost adds up quickly.

- **One-time installation fee per dedicated circuit/order/trunk group:** Sometimes a feature for dedicated toll-free numbers is not actually built on the individual toll-free number, but actually built on the circuit that receives the number. When you are looking at rebuilding or adding features to a dedicated circuit, the fees can be anywhere from $100 to $700 per circuit. The fee may be assigned, not per circuit, but per order or per *trunk group* that is a partitioned section of a dedicated circuit. Be sure to know at what level the fee is assessed.
- **Monthly recurring fee per toll-free number/circuit/trunk group:** Your carrier may charge you a one-time installation and a monthly recurring fee for enhanced features. The monthly fees may quickly add up to more than you are willing to pay for the service. You really need to know what the fee is based upon to know what you will be paying for it every month. It is better to pay $50 per month to set up DNIS on one circuit than to pay $50 per toll-free number for the 500 numbers that ring into the circuit.

- **Release fee when your toll-free number leaves your carrier:** Some carriers get you coming and going in the toll-free world. They not only charge you to set up the new number, but they also charge you when they release the number to your new carrier if you change carriers. There is some work required to release a toll-free number, so carriers that are looking for that extra sliver of profit margin may charge you to release the number as well.
- **Monthly fees to access Web tools:** Carriers that have Web portals may not simply inflate their per-minute charges to provide the service. They may actually recover what they spend in development and maintenance by assessing a monthly fee or per-transaction cost. These fees may not be immediately visible to you until you use the portal for a month or two and determine how much you use and need the service. I suggest that you see how quickly you can cancel the tool if you want to and if there is any contract term on it. If you find the feature is nice, but too expensive, it is better to cancel it in 30 days than to be locked into it for another 10 months.

Chapter 6

Getting the Non-Accountant's Guide to Your Phone Bill

In This Chapter

- Checking out your summary pages
- Determining your best per-minute cost
- Decimal rounding
- Choosing your invoice type
- Identifying common billing issues
- Monitoring dispute progress

Your phone bill is probably wrong. Please, don't take it personally; these things happen. In fact, the problem is so common that some companies exist for the sole purpose of looking at your phone bill and finding the errors so that they can switch you over to a new carrier with a slightly more accurate billing system. Some errors are huge and jump right out at you, and some are small but can cost you thousands of dollars over the course of a year.

These problems plague all carriers, but long-distance carriers seem to have the greatest trouble keeping everything straight. Depending on the complexity of your telecom needs, you may receive phone bills from several different companies, so check them all. In this chapter, I go over the key items to look at in your invoice to ensure you are being billed correctly. I also cover common billing trouble areas, and give you direction on issuing disputes to your carrier to get credits.

Relying on Your Contract

Before you can begin checking anything on your invoice, you need to know what your correct charges should be. Find a copy of your contract for service.

Along with the contract, you should also grab a copy of any amendments your company has signed, any rate notifications that may have landed on your desk, or any correspondence from your carrier that refers to any monthly recurring fees, installation charges, or per-minute costs.

If you don't have all this information handy, you can always contact your carrier and request copies of these files; however, oddly enough, sometimes the special breaks you negotiate don't always make it into your file at the carrier. Funny how that works.

Keep a file of all paperwork and forms you receive from your carrier, especially those that contain any reference to fees or charges. It is better to keep ten pieces of worthless paperwork than lose one that could save you a lot of money every month. If you lose a document that reduces your cost, your carrier may not have it on file either and you will have to start over at square one.

Reviewing the Summary Pages

Every good invoice has a summary section. It is generally on the first or last few pages of your invoice. The section lists aggregate costs for everything on your invoice. The summary is the best place to start reviewing invoices from all of your carriers because it enables you to see the view from 30,000 feet before you focus on each area at ground level.

If you have never reviewed your invoice, you will most likely need a guide. I suggest calling the customer service number on your invoice to find a helpful person to navigate and explain the terrain. Virtually every major carrier has an online presence. You can often answer basic questions by visiting the carrier Web site and looking for a Billing FAQ link.

Regardless of whether you're trekking out on your own, or you have some help, you should start your journey in the summary section. Check out the following sections to find out what kind of information is included in the summary.

Previous balance, payments, and credits

This small section of the summary is a snapshot of your financial state of affairs with your carrier. It is helpful to check this area to confirm that your last payment was received and posted. If your last check was lost in the mail, you immediately see late charges growing on your account. If you don't identify whether your carrier has received your last payment, late charges can build up for several months until you get a call from an accounts receivable representative. After that, you may spend another 30 to 60 days to unraveling the situation.

Here's what you're likely to see:

Your Company Name

Previous Balance:	$1,875.23
Payments/Credits:	$1,800.00
Past Due Balance:	$75.23
Current Charges:	$2,314.57
Late Fee:	$.83
Total Due:	$2,390.63

In this example, the previous invoice was underpaid by $75.23, so a late fee was assessed on the remaining balance. The remaining balance is added onto the new charges that reveal the balance due when the next payment is made.

If you made a payment but it wasn't received, finding out what went wrong may take a bit of time. When the payment is reapplied, you should receive a credit for the late fees. Do not take anyone's word for it that a credit *will* be issued. If you do not see the credit noted on your account summary, it hasn't been posted.

Checking aggregate costs and minutes per phone number or circuit

The summary area of your invoice includes several sections devoted to representing your phone usage in a meaningful way. One part may show all outgoing calls based on the phone number or circuit that originated them; another section may break out all outgoing calls based on whether they were interstate, intrastate, or international; another may lump all outgoing calls together based on the time of day the calls were made. (If you have questions about these call types, please visit Chapter 3 for a full explanation.)

You need to review your contract to see how you are supposed to be billed to determine which summary is the best to use. If you have one rate for in-state calls and another rate for out-of-state calls, a summary by type may be all you need. Divide the total number of minutes by the total charge, and you have your blended per-minute cost for that call type. If the rate is two cents higher than your contract stipulates it should be, you need to do some deeper investigating at the individual call level.

Addressing monthly recurring charges for dedicated circuits

When you activate or deactivate a dedicated circuit, you need to pay close attention to your invoice. If the circuit comes up on August 28, you see a prorated charge for the four days in August that the circuit was active; you're not charged for the entire month.

Understanding that all charges of this nature are prorates is even more important when you are canceling a circuit. You need to confirm that the circuit is end-dated in billing so you receive a prorated charge until the circuit is deactivated. The fee should not appear on future invoices. If the people responsible for processing the cancellation order and physically tearing down the circuit aren't talking to the people responsible for billing the circuit, you could continue to be charged for the circuit.

Understanding installation charges for dedicated circuits

When you activate *(turn up)* a dedicated circuit, you usually are charged for installation. You may see other nonrecurring charges associated with the turn up. Keep your quotes, correspondence, and amendments that you have signed to validate that they correspond with what you are actually charged. These charges should appear on your invoice once and only once.

Try to get the installation fees waived if possible, or order your circuit when the carrier is running a free installation promotion.

Dealing with monthly recurring charges for features and services

This part of your invoice can drive you insane quicker than any other, whether you have dedicated circuits or not. This section lists all the special features you have on your phone lines or circuits. You may or may not even know that you have these features, but you are paying for them. If there is anything in the section that isn't immediately obvious to you, do not feel bashful about asking your carrier for an explanation of the charges and services.

Check your contract to validate what recurring charges you should expect and how they are to be assessed. If you paid $75 for DNIS fees on your dedicated toll-free numbers, you may not remember that the fee was supposed to be per circuit, not per order, per toll-free number, or per location. These tiny distinctions can make a huge difference on your bottom line.

Don't continue to pay the $5.75 per month per phone line for the voicemail service you canceled six month ago; open a dispute and stop the bleeding.

Price quotes for the access portions of dedicated circuits (also called *local loops*) are generally only valid for 30 or 45 days. If you spend two or three months negotiating the finer points of a contract, you may lose the pricing you were quoted. Always get a new quote about the fees for your local loop before you sign a contract for service. The new price may be lower, and you could be pleasantly surprised. It is also possible that your carrier decommissioned some of its hardware and the price jumped up by $150. Better to know that now before you commit your business to a 12-month contract. If you don't know what a local loop is, see Chapter 8. For information about financial issues related to local loops, Chapter 2 is your destination.

Handling installation charges for features and services

When you add services, you probably have to pay a one-time installation fee. Services that go along with toll-free numbers and dedicated circuits are the most common to include installation fees, which can range from $1 per toll-free number to $1,500 per circuit, and everything in between. You need to confirm that all these charges match up with your contract to ensure there hasn't been a breakdown in communication or a typo. This is another section of your invoice where you need your carrier on the phone to discuss what you are paying for and how it matches your quotes. And these charges are nonrecurring. After you pay them, they should go away.

Confirming that all the numbers listed are yours

While you are running down your phone bill, you should confirm that all the phone numbers listed as the source of your outbound calls actually belong to your business. The same goes for toll-free numbers on your invoice as well.

You may think this isn't much of a problem, but number confusion can happen easily. All regular phone numbers and toll-free numbers have about a six-month cooling off period after they're cancelled before they can be reissued. After that, a number can easily be reactivated. If the new customer uses the same long-distance carrier you had with that number, things can get tricky, because your carrier may not have the number listed as blocked or inactive in its billing system. The carrier sees new calls coming onto its network from the number, so the carrier references it to your account and bills you for the usage. Oops. Computer error!

Of course, the carrier isn't in too much of a hurry to fix things. After all, at least *someone's* being charged for the calls. If you see old phone numbers on your invoice, make a list of all the numbers that are not yours and dispute all the usage. Taking swift, assertive action may motivate your carrier to correct the issue quickly.

Analyzing other charges and discounts

If your business is a large telecom customer, you may have negotiated a *volume discount* based upon your usage. Good job! Unfortunately, the down side of this *discount* is the *volume* part. In order to receive the volume discount, you're contractually bound to make a certain number of calls per month.

Usually, you're given a *ramp-up* period of 30 to 90 days so that you can get up and running. Consider this a grace period. After this period ends, you're charged a *shortfall* every month if you don't meet your monthly commitment.

If you have a volume commitment based on the aggregate usage for all of your locations and your carrier doesn't link them together in its billing system, you could see a shortfall charge on the primary billing account. Alternatively, if you meet the volume requirements, make sure that volume discount applies; also make sure the discount corresponds with your contract.

Taking taxes into account

Everyone agrees that taxes need to be paid, but nobody can tell you with 100 percent certainty how much tax should be assessed on your phone bill. By *everyone* and *nobody,* I mean all the people whose job it is to assess and collect taxes for the federal, state, or local government.

There are so many gray areas about what charges apply to what type of service that we're all functioning under a best-efforts environment. Everyone tries hard to do the right thing, and most everyone has a headache. Here are some general statements about taxes:

- Federal taxes apply to interstate and international calls.
- State and local taxes apply to local or intraLATA calls.
- Some taxes are based solely on the usage of a specific type of call. The *Poison control tax* may only be charged on intrastate toll-free calls, but the *new city tax* may only apply to outbound intrastate calls.
- Some taxes are based on a percentage of usage plus other taxes (yes, tax on tax). Local taxes assessed on intrastate outbound calls may be charged based on both the usage and (some or all of) the state taxes assessed for the call.
- In some states, Internet ports aren't taxed, but the dedicated circuit to reach the port is assessed tax.
- . . . And it gets more confusing from there.

Ask your carrier for a list of the taxes it charges customers, as well as the percentage used to assess each tax.

Some cities and states are more heavily taxed, like Los Angeles, California; some, like Las Vegas, Nevada, are taxed less. If you have a high phone bill, shop around for different states where you can set up shop. If your business pays $15,000 or $20,000 in taxes, and you can cut that cost in half, it may make sense to move to the desert, especially if you're in the business of selling air conditioning systems.

Negotiating the Best Makeup for Your Per-Minute Cost

You may be able to save a healthy percentage on your phone bill without even changing the rate your carrier charges. All it takes is to tighten up the accounting. To find this savings, you must first determine the following:

- The minimum duration your carrier charges per call
- The number of digits in the rounding
- Whether you're on a flat or tiered rate plan

The more you pay for your phone service, the more important it is for your carrier to be exact when tabulating charges in these three areas. They may seem like trivial details, but they make a huge difference on what you pay per month.

Understanding call duration and incremental billing

Telecom uses two main call durations: whole-minute increments, or six-second increments. The whole-minute increments are simple and make reconciling your invoice simple. Every call is be rounded up to the next full minute and charged at the rate you agreed to. Because there are no fractions of a minute, there is no rounding, and all of your calls divide out to your exact contracted rate. The downside of whole-minute increments is that most calls do not end at exactly a full minute. For example, a call lasting a minute and three seconds is billed out as a two-minute call. If you're happy paying for 57 seconds of a call you didn't make, then whole-minute increments are for you. If you work for a telemarketing company and most of your calls are 30 seconds long, with each call being rounded up to one minute, you're paying twice the amount that you should.

The best option is negotiating with your carrier for six-second billing increments. This option breaks every call down to tenths of a minute and prevents you from paying for an excessive amount of time that you didn't use. You still have a few seconds on each call that are added, but the net effect is minimal.

Telecom two-digit rounding may not be mathematical rounding

The world of telecom doesn't always use standard mathematical rounding when rating phone calls. If you should be charged $1.011 on a call, your 5th grade math teacher would tell you to round to the nearest penny, so the cost is $1.01. The reality of it is that many telecom companies with two-digit billing charge $1.02 for that call.

Here's the rationale for increasing what you pay by a penny: No one can expect you to pay a half of a penny for a call, and so why not simply round the fee to the next penny?

If you only make five calls a month, this creative math is not a problem. If you make 500,000 calls a month, however, the pennies add up quickly. If you cannot get four-decimal rounding from your carrier, at least negotiate standard mathematical rounding until you can find a carrier that will give you four-decimal rounding.

Beware of the lure of billing units. Some carriers do not charge you in increments of seconds or minutes, but in billing units that do not correlate easily into reality. If your salesman cannot explain the rates of your calls in one sentence, avoid it.

One-second billing units aren't necessarily better than six-second units. It does seem logical that the more accurate your duration is recorded, the more accurate your invoice will be. That is, unfortunately, only half of the equation. In the end, your rate also depends on your decimal rounding. See the following section, "Dealing with decimal rounding."

Dealing with decimal rounding

If, as in the previous section, "Understanding call duration and incremental billing," you run a telemarketing business that makes many short phone calls, you may want to negotiate a six-second billing plan so that your invoice is more manageable.

But wait — there's more. You need *four-decimal rounding.* Many companies in America have two-decimal rounding — every call is individually rated to the nearest penny. If you have a dedicated circuit with a rate of three cents per minute on calls and you make a six-second call, you should legitimately be charged $0.003 — three-tenths of a cent — for the call. If you have only two-digit rounding, you're going to be charged a penny for the call. This does not seem like much, but $0.01 for a0.1-minute duration call gives you a per-minute rate of 10 cents.

If most of your calls are of short duration, two-digit rounding can easily raise the average rate you pay. Consider three- or four-digit rounding if you have the option. If your calls are rounded to three decimals in a six-second call at three cents a minute, you're charged the correct cost at three-tenths of a penny. Sounds good? The only problem with *that* is that most long-distance rates are not in full-penny increments. Competition has pushed rates down so that your contract for dedicated circuits may have a rate of 2.9 or 2.2 cents per minute. When you're dealing with 300,000 minutes per month, the difference in your phone bill between the two rates is $2,100. The only way to effectively capture your rate of 2.9 cents per minute is to have every call rated to the fourth decimal point. In this case, a 12-second call (0.2 minutes) costs $0.0058, and your per-minute rate is intact. If this amount is rounded to two decimal points, you pay about double for the same call. If you have a small number of calls per month, and they are all of long duration, the digits in rounding don't affect your bill too much. But if you have a high volume of calls under a minute, the cost savings through this one change could be 50 percent.

The savings you see don't increase if you secure five- or six-digit rounding. At that level, the amount of money is so small that it would take hundreds of thousands of calls to add up to a dime of savings. If a carrier sales rep claims that the carrier's billing plan will save you money because it has five-digit rounding, have the rep crunch the numbers to prove it to you.

Figuring out flat and tiered rates

Setting call durations to full-minute or six-second increments, as well as ensuring that call rates are rounded to four digits can make a huge difference to your phone bill, without technically changing the rate you pay per minute. Well, you can negotiate your per-minute rate, as well. In the realm of call rates, there are two methods: the flat, and the tiered.

- **Flat rates:** Simple call rates (also called flat rates) break your calls down by type (intraLATA, interLATA, intrastate, interstate, and international), and charge you a set rate for each category. The breakdown may even be simpler, with one set rate for your in-state calls, and a different rate for your out-of-state calls. If you have a flat rate of six cents per minute for all in-state calls and three cents for all calls that terminate out of state, you have nothing to confuse you. Whether you call a cellphone, a phone number that belongs to the local Bell carrier in your area, a VoIP phone or any other type of phone in your state, you're charged six cents per minute. This system makes life easy, but you may be paying more than you need to for these calls. Flat rates are relatively uncontroversial.
- **Tiered rates:** Complex rates (also called tiered rates) are becoming more and more common with carriers. Competition has forced the per-minute rates for calls down, so the carriers have reevaluated how they charge for calls. Every long-distance carrier has a contract with every local carrier in the U.S. It may cost seven cents per minute to terminate a call to a cellphone company, but it may cost one cent per minute to complete to your local Bell carrier (Pacific Bell, Bell Atlantic, Bell South), and maybe four cents to complete to a company that provides VoIP service. Long-distance carriers now tier their prices accordingly. If most of your traffic terminates to the local carriers with better rates, you could save some money by going to a tiered rate plan. Conversely, if most of your calls end up going to cellphone providers or *competing local carriers* (CLECs) like MPower or Vonage, your costs could go up substantially. Deciding on a tiered plan can be an arduous task. The following sections give you more information to make your decision a bit easier.

Always check over your first invoice from any new carrier with a fine-toothed comb. You may have spent a lot of time negotiating a better rate for an area you call frequently, only to discover that the rate was never sent on to the billing department.

Why tiered rates have become popular so fast (if you care)

Telecom customers have become very sophisticated in the past ten years. Some customers have advanced routing tables that look at every outgoing call and send it out over the cheapest route. This idea may sound great, but as these new, technically advanced customers have entered the market, they exploit any carrier they can find with a flat-rate pricing schedule. As carriers began charging anywhere between 3 and 9 cents to complete calls to cellphones or *competing local carriers (CLECs),* these uber-customers routed those calls to carriers with 2½ cent per-minute flat rates. After a few months, the flat-rate carrier would see a dramatic spike in traffic to the expensive terminations and would have to correct the problem or face bankruptcy — and so all the carriers eventually created their own tiered rate plans to stay alive.

Knowing your tiered plan do's and don'ts

Tiered rates sound great in theory, but the devil's always in the details. Before you sign on the dotted line, consider the following do's and don'ts:

- **Don't get a tiered plan if you call a lot cellphones or CLECs.** Tiered pricing contracts commonly require users to pay a premium to terminate calls to cellphones and competing local carriers.
- **Do get a tiered plan if you call a lot of the local Bells.** Tiered plans commonly require users to have a high percentage of calls terminate to the local Bell carriers like Bell Atlantic, Nynex, Bell South, Qwest (local), and Pacific Bell. Great, right? Well . . . read the next item in this list.
- **Don't get a tiered plan if you don't call *enough* local Bells.** If you terminate fewer calls to these carriers than the required percentage (which varies from around 65 to 80 percent), you may face a substantial per-minute or per-call penalty.

Do review your calls before you make a decision. Before you agree to a tiered pricing plan for your company, make sure that you look at all of your outgoing calls for a few months. Better yet, ask your carrier to analyze your traffic and determine what the net result will be. Ask for a report showing the percentage of calls that fall into each tier, and your aggregate cost per minute. It would be horrible to go into a tiered pricing contract, only to find out three months later that your average cost per minute went up five cents.

- **Do know the intrastate rates on calls.** Around the year 2000, carriers began to notice that their customers with dedicated circuits were inputting origination phone numbers on their calls. If a telemarketing company in Texas was working a project for a customer in Florida, they would list the client's Florida phone number as their caller ID. This enabled the people who received the telemarketing calls to contact the Florida business that hired the telemarketing company in Texas. A good idea, really — it made the Florida business look really snazzy. The only problem was that the local carrier in Florida saw the calls as having Florida origination numbers, and Florida termination numbers. In the eyes of Bell South, that made it an intrastate call, so they charged the carrier 14 cents per minute to terminate the call. Usually, interstate calls cost two cents per minute. As the carriers realized they were losing money, they put measures in place to pass on the cost to their customers.

  The moral of this story is that if your business has a direct contract with a carrier and your phone system inputs a Caller ID number on your dedicated circuit that does not represent your business's physical location, you need to check your phone bill. If you see summary sections for intrastate calls in areas where your business cannot originate a call, you need to take swift action to correct your billing. Your carrier may have another billing plan available, or you may need to find another carrier altogether.

Introducing OCN, the Rosetta Stone for tiers

Every carrier has its own system of grouping the local carriers into tiers. One carrier may place Verizon in tier 3, while another may place them in tier 1. Some carriers also assign tiers to each LATA, and some assign tiers only at state level. One unifying thread enables you to make sense in a world that lacks uniformity.

An *Operating Company Number,* or *OCN,* is a four-digit number that the telecom world uses to identify a local carrier. Your carrier can provide you with a list of OCNs and their associated tiers. Of course, this information still doesn't tell you how much money placing a call to a number in Maine at 207-224-XXXX will cost. To determine per-minute costs, look at the OCN tier list provided by your carrier; find the OCN of the local carrier that provides service to the number you're dialing and identify the tier of the call. Then you can determine the LATA of the same phone number, and then you are ready to dive into the LATA-tier price list provided by your carrier. Clear as mud?

Your carrier should be able to provide a list of every area code and prefix (that's the first six digits in any domestic phone number) in the U.S., along with their associated OCNs. If you have a large analysis you are crunching, you need this information in a database or a text format.

The OCN-to-tier list that your carrier gives you doesn't include all the OCNs in the U.S. If an OCN doesn't appear on a carrier's list of tiers, the general rule is that the OCN is in the most expensive tier. Some carriers offer a disclaimer to this effect.

If you want to spot-check a few rates, you can visit a Web site to check individual phone numbers. Follow these instructions:

1. **Visit the Local Calling Guide at the Outsider's Report Web site.**

   The Web address is `www.localcallingguide.com`. The Outsider's Report is a Web site that has provided telecom information in one form or another since 1994.

2. **Scroll down to the middle of the page and click the Search `AreaCode/Prefix/OCN` link.**

   The Area Code/Prefix Search page appears.

3. **Enter the area code in the text box marked NPA. Enter the first three digits of the phone number into the text box marked NXX.**

   You can leave the other boxes blank.

4. **Click the Submit button.**

   The next screen lists all the information you need. The four-digit OCN is cramped right before the name of the local carrier, but is easily visible.

5. **Write down the OCN and LATA for the phone number.**

   Be sure to cross-reference the OCN to the appropriate carrier tier with the spreadsheet your carrier provided.

6. **Write down the tier for the OCN and identify the correct rate per minute you should be assessed.**

   Quickly dividing the cost of the call by the total number of minutes gives you the contracted rate. If the numbers don't match, call your carrier to determine why.

Receiving Your Invoice Your Way

You can receive your invoice in any number of ways, and as your company and your phone bill grow, you need to ask for your invoice your way. Most carriers have at least two options for billing, and sometimes more.

Going old school with paper invoices

Most people and many businesses still receive their phone bill in the form of a *paper invoice* that arrives in the mail with a remit envelope enclosed. The bulk of your invoice consists of the itemization of every call made during the invoice period. The itemization includes

- The time and date every call was made
- The phone number that received the call
- The duration of the call
- The cost of the call

The lines of data that describe the specifics of your calls are referred to in the telecom world as your *call detail record*, or *CDR*.

Having the best of both worlds with CD-ROM billing

When your phone bill has so many pages to it that it is thicker than an unabridged dictionary, you should ask to have the section of your invoice that individually lists your phone calls removed from the printed invoice and sent to you on a CD-ROM. This format is better for large companies because you can import the call detail from the CD-ROM to a Microsoft Excel spreadsheet or Access database. Plus, it saves paper.

After you've entered all the call information into a database, you can manipulate the calls to find a cost per minute on each call, sort the calls by area code, cost, date, time, or view the information however you choose.

CD-ROM billing isn't a free service with all carriers. The fee is generally not that expensive, and well worth the benefit of having all the data of your phone bill at your fingertips.

Getting digital with online billing

Many new carriers have elaborate Web sites that enable you to sign up for new services, make changes to your account, and see your invoice. If you're a techie-minded person, this may be your dream come true. Some carriers even have real-time invoice posting so you can see what your balance is, including the call you just finished making two minutes ago. If this is something that excites you, you can easily find out whether carriers offer such services by checking the Web. Printing your invoice, or at least a summary page, allows you to keep a record for posterity. Many companies provide PDF versions of your invoice to download and save on your computer as well.

Knowing about Billing Issues

No billing system in the world is infallible. A billing system isn't just the computer that rates your calls and prints up your invoice. If you want to account for every variable, you have to include every switch in your carrier's network that logs in calls and send them to be billed, as well as all the switches of the companies your carrier contracts with around the world that do the same thing. That's not so easy. With that said, here are some specific annoyances you should prepare for that won't be very visible in the summary section of your invoices.

Keeping an eye out for long-duration calls

Once in every ten million calls or so, a call doesn't hang up just right. You called a client and spoke to her for ten minutes, but the switch that recorded the duration of your call never realized that you hung up. The industry refers to calls that don't complete properly as *hung calls.* They are generally very easy to identify because they last, not hundreds, but actually thousands of minutes.

Some legitimate calls can last 400 or 500 minutes, but any long-duration call needs to be investigated internally before you dispute it with your carrier. Here are some legitimately long calls:

- The line may be used to transmit data for days on end with a dialup modem.
- You may have a problem with your automated phone system or calling card platform that actually keeps calls open for thousands of minutes. Unfortunately, that's not a hung call — it's your responsibility to both pay for the call and fix the problem with your phone system.

The easiest way to tell a real, carrier hung call from a call that was legitimately open for 1,000 minute or more is to check how many suspiciously long calls you're billed for in a six-month period. If you have two hung calls in the same month, the problem is probably with your phone system, or the phone system of whomever you are calling. It is statistically improbable that the one-in-ten-million event will occur to you twice in the same month. You need have your hardware vendor investigate your long-call phenomenon.

If you notice long-duration calls on your phone bill to people you don't know, this is a red flag that you may have fraud on your phone line. If you have a very basic phone system, or none at all, someone may be tapping into your line directly. If you have a dedicated circuit and a phone system with every feature known to man, your whole system may be compromised. If this is a concern for you, please check out the fraud section in Chapter 2.

Making note of duplicate calls

Duplicate calls are exactly what they sound like they are. The exact same call is billed to your business twice. It takes a bit of skill to find out whether a duplicate call is a legitimate charge or not, because most carriers have a checking system to route out duplicate calls before they make it onto your invoice. In order to have legitimate duplicate calls you must have two or more calls that share exact matches on all the following:

- Date of call
- Start time of call
- Phone number originating the call
- Phone number called
- Duration of call

Just because you have two calls to the same number with the same start time does not guarantee that you have duplicate calls, especially if both calls are less than a minute in duration. Start times on calls rarely show the seconds, so it's technically possible to make two calls of eight seconds in length to the same phone number within the same minute.

The calls look like duplicates, but really, your employee got the voicemail of the person he intended to call, hung up, and absently pressed the redial button when he meant to press another button on his phone. Realizing his mistake, he hung up again when he received the voicemail a second time. Presto! You have two legitimate calls in the same minute.

Genuine duplicates usually appear in bulk. You generally won't see only one call affected, but an entire section of calls. Duplicates usually occur within a date range or even within a few hours of the day, and you may see as many as 50 percent of the calls duplicated. You can easily identify duplicates because calls of all lengths are affected. The moment you look at your bill and see two calls that started at the same time, and each lasted an unusual length of time (35.4 minutes, for instance), you know you have probably found a legitimate problem. You should issue a dispute with your carrier.

Handling overlapping calls

Overlapping calls are like duplicate calls in the fact that they can be legitimate. If you notice that one of your phone numbers has a call that started at 12:10 p.m. and lasted five minutes, and the same phone number made another call at 12:11 p.m. that lasted three minutes, you could have a problem.

Your carrier will probably assert that you have three-way calling and that the person who made both calls simply made a conference call. If that is true, please pay your invoice and chalk up another important learning experience. If your employees generally don't use conference calling, you need to issue a dispute and have the carrier look into the calls a little deeper.

Almost every phone in America has the ability to make a three-way conference call. If you ever hang up one call and immediately dial out again, you may hear a special dial tone that is broken at the beginning called a *stutter dial tone*. This is your local carrier alerting you to the fact that the next call you make can be linked up to the last call you just finished. If you see overlapping calls on your invoice and they are legitimate, remember the sound of the stutter dial tone. If you hear it, hang up and wait three seconds before you dial again. After a few seconds, you hear the normal dial tone and you have no chance for overlapping calls.

Watching for connect signal issues (especially on international calls)

A *connect signal* is a message sent through your carrier's network when your call is answered. A live person doesn't have to pick up the phone for the call to be answered; any answering machine or voicemail system will do. The moment the receiving phone accepts the call, the connect signal is sent, and when the signal is received by the network, your carrier begins tracking the duration of your call so that you can be charged for it. Some carriers don't begin billing when the connect signal is sent through the network, but instead begin the clock as soon as you dial into the network. On a dedicated call, you can begin to incur per-minute charges even if no one answers the call. If your carrier seems to be charging your business for calls the instant you enter into its network, you need to speak to your carrier and see whether you can change your platform or move to a part of the network that bills your business only at time of connection.

Incorrectly starting the duration clock on a call is more common on international calls where your carrier has little leverage to correct or regulate when the clock begins billing. Unfortunately, if your carrier cannot force the case with whatever carrier it uses to make international calls, you have two chances of receiving a credit for the 30 seconds extra: slim and none. This situation is only visible when you receive your phone bill a month after frantically trying to reach someone overseas whose phone system is down (maybe there was a big storm or some other natural disaster that affected phone systems). If you're billed for the 15 calls you made for 3 days when the phone lines were down in that country, you know that the phone company has been skimming a little off the top.

Facing facts about old calls

Finding calls on your phone bill from a few days prior to your billing period is pretty common. In fact, this is a normal occurrence, and usually you're just charged for the calls once. You can check your previous invoice and look for the specific calls to be sure, however. If the exact calls also appear on your previous invoice, dispute the charges.

Understanding why cross talk causes strange billing

Cross talk is the industry term for that weird moment when you're on a phone call and suddenly hear another conversation on the line. Frequently, the other people cannot hear you, even though you can hear them. This is the only way you can legitimately have someone else's phone calls on your invoice without fraud being the cause. Cross talk from your carrier is a result of the insulation around the wires on your phone line becoming thin, or missing completely, and touching the bare copper wires of another phone line. Along with hearing the conversations of others, your phone may also ring even though no one is calling you.

What is essentially happening is you are (or the other person is) receiving the dial tone transmitted for the other person's phone line when you make an outbound call. The dial tone you hear is attributed to a specific phone number within your local carrier, so by accessing someone else's dial tone, the local carrier bills the call to the other phone number. If the cross talk gives someone else access to your phone line, your long-distance carrier won't be able to prove it. Your carrier sees this call as a normal, legitimate call from your phone. Take care of this problem immediately with your carrier because it skews everything in the billing world. Cross talk is very rare and your carrier will not give you a credit for the calls unless you can reference a trouble ticket you opened with their repair department for the problem.

Every evening at a predetermined time, every routing computer in every carrier network sends data for completed calls in the past 24 hours to a billing computer. Domestically, this happens with monotonous regularity. If you have calls that go to the Caribbean, Guam, or to another international destination, another day or two may pass before the record of your call is finally sent to the billing center. This delay in relaying the call detail is what causes old calls from up to seven days before your billing cycle to appear on your invoice.

According to U.S. federal law, your carrier can't legally invoice you for calls older than 90 days. If your carrier forgets to invoice you for calls on a new circuit or toll-free number that you have owned for 6 months, the carrier can only legally back-bill you for 90 days. Even if you receive an invoice with six months of old calls on it that you have never seen before, you only have to pay for the last three months. This three-month shelf life is known in the industry, but because the carrier has nothing to lose by asking you to pay for all the back traffic, the billing department will do what it can to get as much money from you as it can.

Handling Your Billing Disputes

When you review your invoice, you may find a few issues that don't add up right. Maybe that rate reduction for calls to Canada you negotiated didn't take effect last month like it was supposed to, or perhaps your carrier is still billing you for the dedicated circuit you cancelled four months ago. Before you rush to issue a dispute with your carrier, you need to know what, and how, you can dispute.

Understanding your carrier's process

Every carrier has its own process for addressing billing disputes. Before filing a dispute, find out the procedure. Before submitting the dispute on the correct form and to the right e-mail address, ask about the timeline for resolution. The standard turnaround time on billing disputes is 30 to 60 days. I would like to say this is a clean and efficient process, but I have seen the process take over six months.

You can call a customer service representative at your carrier, but often all you need to do is check your paperwork. Every contract for long-distance service has a section regarding billing disputes. If you don't have a contract for phone service, you can usually find a similar section embedded within your phone bill. And the general policy is often listed online, as well. Regardless of where it lives, the section tells you all the logistics of the dispute process. In particular, the information covered includes the following:

- **The method for issuing disputes:** Some carriers require you to submit a dispute in writing, while others begin working on a dispute from a phone call. Your contract may also stipulate the required supporting documentation you need to supply before the carrier begins working on your issue. Only after you supply the information can you get your dispute into the system and get your credit rolling.
- **A time frame for you to dispute charges:** As a general rule, carriers only allow a 30- to 45-day window to dispute any charges on your invoice. After that time frame, the carrier can (and will) summarily reject any dispute you submit, regardless of its validity.
- **Info on whether you must pay a disputed charge:** Your contract stipulates whether you must pay disputed charges while the amount is in dispute. This is an important detail, because you will be charged late fees on the disputed portion of the bill if you withhold money when you're not authorized to do so.

Working the appeals process

If your dispute is legitimate and has been denied for some unfounded reason, resubmit the dispute. The carrier surely has an appeals process, so if you feel unjustly rejected, restate your case. Accepting the fact that some of your disputes will be rejected is healthy, because you may be unaware of contract details hammered out, or overlooked by, your boss. Graciously understanding the disputes that are denied also generates a good rapport with the dispute department and ensures that credits flow smoother when they are granted.

No matter what your carrier requires, however, you should always send disputes in writing. You don't have to use fancy calligraphy and secure the envelope with a wax seal, but you need a paper trail if the dispute ever gets to the point of mediation. When you document the dispute, listing as much information as possible is important; detailed information can prevent delays, and, of course, delays just perpetuate problems.

Instead of writing a letter of dispute, you could call your representative and say that you think your interstate rates to Montana are wrong; the carrier may make a good-faith effort to fix the problem, checking all your outbound calls to Montana. Two or three months later, the rep may call you back and say that he or she couldn't find anything wrong. That's when you review your notes and say, that it was toll-free calls *from* Montana that have the wrong rate. Get all the information in one place, in writing, the first time to prevent a mountain of frustration later.

Tracking your disputes

Of course, I hope you don't have multiple disputes with your carrier (if you do, you need to start shopping for a new carrier), but it could happen, especially if you have complicated services and a large phone system. If you have more than one dispute in process, you need to track them carefully. The carrier may aggregate disputes for the same issues that appear over several invoices, or it may lump them together by month. The carrier may also lose disputes that you submit. In the end, it is your business's money at stake, so I don't suggest you rely on someone else to manage it.

Use Excel Workbook to track your disputes for the year. Make the first tab a summary by month with cells aggregating the total dollar amount for disputes by month. Itemize all of your disputes chronologically on the second tab. The columns of information on the second tab should include:

- **Invoice date for the dispute:** If you dispute rates on the July invoice and your carrier checks the June invoice, well . . . that's not very good.
- **Dollar amount in dispute:** The carrier needs to know the bottom line you are expecting as a credit. Determine this amount on your own — don't wait for your carrier to calculate it for you.
- **Specifics of the dispute:** In a sentence or two, give the gist of the dispute for easy reference. For example, you might say, "Acme Corp. was charged 14 cents a minute in July for outbound calls to Canada even though the contract stipulates that all outbound calls to Canada will be charged at a rate of 8 cents a minute."

  Keep this info short and simple. You have other documentation if you really need to go into specifics, but on a tracking spreadsheet, you simply need enough to jog your memory.
- **The date the dispute was submitted:** Having this information in a separate tracking document validates that you submitted the dispute within the time frame required by your contract.
- **Whether the dispute has been approved, denied, or remains pending:** You need to track the results of the process so you know which issues are still open.
- **Credit amount, if approved:** A full credit is not always given, so you need to track the actual credit amounts along with the amount in dispute. Having this information in one place reminds you that you need to pay the remainder that was not credited.

Most carriers are relatively reasonable, even if the process is a bit arduous. If your dispute is valid, you should receive a credit. If your dispute is valid but it goes back farther than 30 days, your carrier has the discretion to ignore the claim. Carriers often offer courtesy credits for issues that may not have been their problem. They want to hang on to their business, and the best way to do so is to keep customers happy. Keep this information in mind and you will appreciate it all the more when a carrier bends the rules for you and credits your account for the past six months; likewise, you will understand all the better why sometimes a carrier has to hold firm and deny your claim.

Part III
Ordering and Setting Up Telecom Service

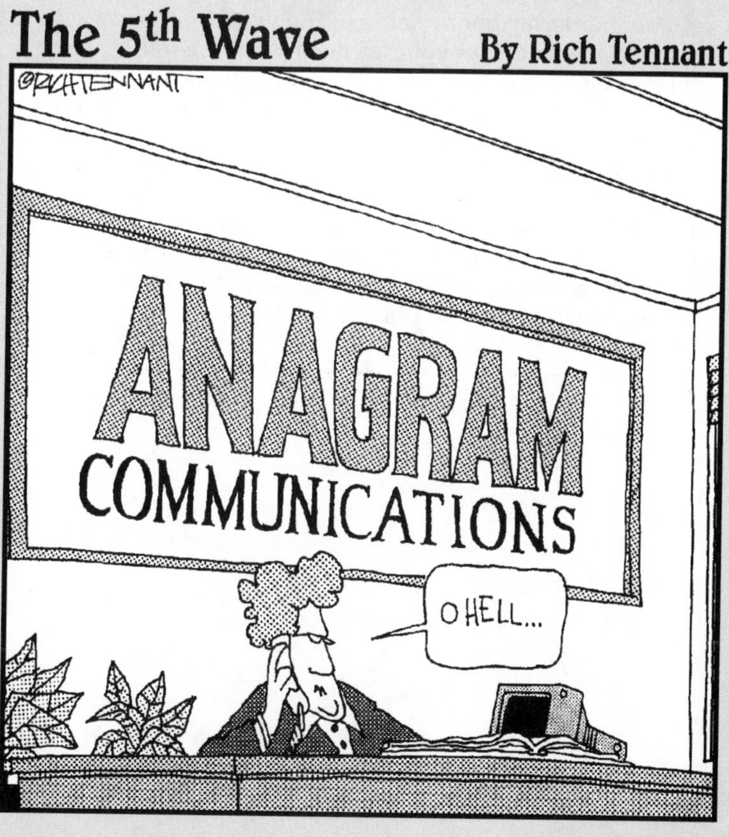

In this part . . .

In Part III, you discover everything you need to know about ordering phone service. You may be ordering another phone line for your fax machine, a toll-free number for your customer service department, or a dedicated circuit for your new office, but no matter what you want to do, all of the information you need is in this part. The provisioning system for all of these items could come to a grinding halt for a number of reasons, so you need to know how to resolve any delays and get those phones up and ringing. The bonus round of this part is the Chapter 10, which covers the steps necessary to activate dedicated phone service and toll-free numbers.

Chapter 7

Ordering Regular Phone Lines and New Long-Distance Service

In This Chapter

- Ordering regular phone lines
- Changing your long-distance carrier
- Rejects and their reasons
- Validating your long-distance carrier
- Casual dialing with 10-10 codes
- Local number portability

Most of us don't order new phone lines every month. They are generally a lot like pieces of luggage in that way: you get some, you use them, and you only buy more when it is painfully obvious that your current supply isn't cutting it. This isn't very helpful because by the time you need more, you will probably have forgotten what you did to order them the last time.

To avoid that stress, you need to understand how long it takes to activate a new phone number, add or change the long-distance carrier on it, validate the long-distance carrier you have on your lines, and cancel lines or service when you're done with them. This chapter guides you through the steps necessary to do all of that, informs you about how to avoid pitfalls during the process, and tells some other fun things you can do with long-distance carries. If you have concerns that you have too many — or not enough — phone lines, please take a trip to Chapter 2 for information on how to determine what you need; otherwise, you're free to forge ahead in this chapter.

Ordering a Phone Line

The world of telecom has two types of service, switched and dedicated. *Switched* service is what you probably have at your house. If you don't have a phone system, but just have a jack in the wall and a phone number that is assigned to it, you have switched service. These are the regular phone lines (sometimes called POTS, or plain old telephone service) you use every day. Switched service is so commonplace that when you think about phone lines, you think about POTS.

Ordering a switched phone line seems like a simple task. All you need to do is pick up the phone and call your carrier to order one, right? Well, not exactly. Not every carrier can give you a phone number.

Need a line? Calling a LEC

You can only order a phone line from a local exchange carrier (LEC) like Qwest, Bell South, or Verizon; competing local exchange carrier (CLEC) such as Mpower; or a cellphone company. These companies have been authorized by the U.S. government to distribute phone numbers. You can't order a phone line from your long-distance carrier because the long-distance divisions of MCI, AT&T and Sprint are only authorized to complete calls sent to them. The U.S. government has not given them the authority to deal in the collection or distribution of phone numbers for their customers.

Calling your local carrier

If you have any phone service at all, you receive an invoice from your local carrier every month for it. Having local services makes life easy when you want to order more phone lines because you just have to call the customer service number listed on your invoice and order them.

If you have just paid your local phone bill and don't have one handy, don't worry, there is still a way to find their phone number. If you have a regular local carrier, like Qwest, Bell South, or Verizon, you can simply dial 611 to reach customer service. Usually, this number is used for reporting trouble on your line, but many local carriers have adapted it for general use. If you call 611 and don't reach an automated attendant system, asking you to "please select from the following menu options" but instead reach a live human, the rep who answers the phone will be more than happy to transfer you to customer service.

If you do not have a standard local carrier like Bell South or Verizon, but instead have a CLEC, you may hear any variety of recordings or responses. In this case, you have about five minutes more research before you can call them. Go to the next section and walk through the steps to find and contact your local carrier.

Getting local versions of long-distance companies

MCI, AT&T, and Sprint do have divisions that provide local service that can give you a phone line. You would call MCI local, AT&T local, or the local division of Sprint in Las Vegas, Nevada. This being said, unless your business has been targeted by one of these companies, or you live in Las Vegas, these companies probably can't provide you with anything other than long-distance service.

Finding out who your local carrier is

If you don't have a phone bill handy, and, for some reason, dialing 611 doesn't work, you can always dial the operator by pressing 0. Listen carefully, because the operator will probably identify the name of your local carrier. The only problem with this technique is that some local carriers contract out operator services, so instead of reaching a local operator in your city, you may be transferred to someone in the Philippines who works for a company that provides operator service to many local carriers. It's possible that your operator has no idea what the name of your local carrier is.

Don't worry. If all else fails, there is a foolproof way to get a phone number for your local carrier; it's called the Internet. This is a simple process after you connect to the Internet:

1. **Visit `www.localcallingguide.com`.**
2. **Scroll down to the middle of the page and click the Search Area `Code/Prefix/OCN` link.**

   The Area Code/Prefix Search page appears.
3. **Type your company's area code in the box marked NPA. Type the first three digits of your company's phone number in to the box marked NXX.**
4. **Click the Submit button.**

   The Search page refreshes and lists your company's city, state, and local phone carrier.

When you know the name of your local carrier, you can find a phone number for it by starting a search from `www.google.com` or `www.yahoo.com`. When you visit your local carrier's home page, you won't have to look too hard to find a Contact Us section and a phone number for customer service.

If you move to another town, or across the country, you probably won't know who the local carrier is for your new house or business. No problem. All you need is a phone number for any home or business within a five-mile radius of your new location. Plug the area code and prefix into the `www.localcallingguide.com` site and you can find out their local carrier. Because local carriers generally don't change within the same town, you have all the information you need to order your new phone lines.

Placing the order

When you call your local carrier to place an order for new service, you have to navigatge through the maze of prompts. Eventually, you reach a nice person whose job it is to take your order. If you have existing phone service, this is generally a painless procedure. You're asked to confirm your identity, your address, and the billing name on your invoice. When the preliminaries are over, you're asked how many new phone lines you need, and then to validate the address, the features, configuration (is the new number going to be in a hunt group with your other lines or not?), long-distance carrier, and area of coverage for your long-distance carrier. If you're adding to existing service, you should tell the agent to duplicate what you have on your existing lines.

Securing your phone lines with a PIC freeze

A *PIC (primary interexchange carrier) freeze* is a logistical security device. If you tell your local carrier to place a PIC freeze on your phone lines, the carrier locks your long-distance carrier on your lines and never changes it — until you personally call your local carrier, identify yourself, and have the freeze removed.

Depending on the procedures your local carrier has set up for activating and removing PIC freezes, you many be able to simply call your local carrier, or you may have to fax the carrier an authorization. After the PIC freeze is removed, your new long-distance carrier can send the local carrier another order to convert your lines. I am sure you're wondering why you would want to go through all this effort whenever you want to change long-distance carriers. Glad you asked.

The benefit of a PIC freeze is that it protects your lines from having the long-distance carrier changed on it without your knowledge. In the industry, *slamming* occurs when a customer gets a new long-distance carrier that was never requested. PIC freezes are generally available for free from your local carrier, and automatically reject any attempt to change the long-distance carrier assigned to the protected phone lines. Carriers don't offer account-level PIC freezes, so your local carrier has to place a freeze on each individual phone line you have to ensure that they are all protected.

You may have to place PIC freezes on both intraLATA and interLATA lines, because carriers usually handle them seperately. If you place a freeze on only interLATA long-distance lines, you could have a long-distance carrier slam your intraLATA calls. Getting the freeze is fast, free, and easy, so place a freeze on all of your lines, and both the intraLATA and the interLATA options.

Understanding and recovering from being slammed

When your phone line is slammed, it is not always done intentionally. Quite often, it is the result of a typo on the part of someone at the carrier or one of the carrier's customers. Many companies have so many phone lines that they don't have a solid accounting of them. They may have 20 numbers in sequence and think their last number ends at -2328. In reality, the group of numbers may end at -2327. If the long distance carrier sends an order to your local carrier for the complete range of numbers, including the extra number (-2328), the actual owner of the phone number receives an order for a new long-distance carrier.

A similar typo could be made by an eyesore data entry employee in the provisioning department at a carrier where the number -2341 was entered rather than -2431. The advent of antislamming laws in the U.S. has created a large disincentive for unscrupulous long-distance companies who used these tactics to bring on new customers in the past. Now that sufficient legislation is in place, the percentage of people intentionally slammed is very small, with the vast majority being the unintended result of human error somewhere in the system.

Understanding your long-distance options

Your local carrier identifies two areas of coverage available for your long-distance carrier: intraLATA and interLATA. *IntraLATA calls*, also known as *local toll calls* or *Zone 3* calls, cover any phone number you dial that is over (roughly) 13 miles away but still located within your local access transport area (LATA). InterLATA calls cover any calls you make to phone numbers outside your LATA. It is usually a better option to have your long-distance carrier pick up both areas of traffic, because your local carrier generally charges for the intraLATA calls based on mileage. In other words, if you call to a phone number 100 miles away, but the number is still within your LATA, your local carrier may charge you more than it would if you called someone 20 miles away. As a general rule, long-distance carriers charge you a flat per-minute rate, regardless of where you call in your LATA. If you don't make these calls often, how you're charged may not be much of an issue. Check your last phone bill to see what your average rate per minute is for these calls, and then decide which carrier is better. If all this talk about LATAs is making your head spin because you don't know what one is, please jump over to Chapter 3 for a crash course.

You probably won't simply wake up one day with an epiphany from the universe telling you that your long-distance carrier has changed. Most people figure it out by either receiving an invoice from a new long-distance carrier or discovering that all of their long-distance calls are suddenly blocked. As soon as you realize something is drastically wrong, call your local carrier. When you have a customer service agent on the phone, you should do the following:

1. **Find out the name and contact number for the carrier that slammed you.**

   This information is necessary so that you can call the offending carrier and research what happened.

2. **Switch your long-distance back to your old carrier.**

   Your local carrier is the only company that can do this, but it doesn't keep a record of previous long-distance carriers on phone lines. If you had a mainstream carrier like Sprint or MCI, tell your local carrier to switch your phone lines back and you won't have any more problems. If you used a reseller or lesser-known carrier, you may need to call your old long-distance carrier to send in another order to switch your lines back.

3. **After your phone lines are switched back, place a PIC freeze on all your lines.**

   Just because it has already happened once doesn't mean it can't happen again. You don't need the hassle — *again.*

4. **Dispute all charges for calls that went across slammed lines with your local carrier.**

   In short order, the second you start the dispute process, you will probably be chatting with a representative of the carrier that slammed you. This conversation will either occur because the carrier is responding to your dispute of the charges, or because you directly tracked the carrier down to discuss the matter. Unless the new long-distance carrier can confirm (by confirm I mean *prove*) that you, or some authorized person at your organization, requested that your phone lines be changed, the slammer should credit your charges for long-distance service.

   The cost of paying for slamming fees for violating FCC regulations is more than a few dollars, so it is in the best interest of the offending carrier to credit any charges and make the problem go away. If you have 1,000 minutes of usage to Dussledorf, Germany, or some other overseas location, you may hear some resistance when you ask for a credit. The truth of the matter is that you did make the calls, and they did complete them. A strong argument could be made that they should receive something for the service they provided. In this case, the very minimum you should negotiate from them is to receive a credit to adjust the rate they charged to what you would have paid on your old carrier.

5. **Request a credit for the change charges.**

   Your local carrier is going to charge you a few dollars every time it changes your long-distance carrier. If the fee is $4.50 for every carrier change, you may be hit with an $18.00 charge *per line.* That's because you have to pay four times; once to switch for your intraLATA calls to the new carrier ($4.50), once to switch your interLATA calls to the new carrier (another $4.50), two more times for moving them both back (another $9.00). If you have ten lines or more that were slammed, you should demand a credit for the change charges.

Planning for inside wiring

Your local carrier is only responsible for bringing your phone lines to the *minimum point of entry* or *MPOE* (pronounced *em-poh*). If the main phone closet is at one end of your building and the new phone lines for your fax machines need to be run to the other end of it, someone needs to connect the phone lines from the MPOE to the room where your fax machines are sitting. The section of wiring that runs from the MPOE to your phone system or telephone is called your *inside wiring*. Your local carrier has a set policy that stipulates whether it provides inside wiring, and how much installing inside wiring costs. The local carrier may be more than happy to provide the inside wiring, or it may have a blanket policy to only drop phone lines at the MPOE.

If you are a thrifty shopper, you should call your hardware vendor to install the inside wiring and not use your local carrier. Your hardware vendor is probably cheaper and more flexible, and it may be willing to do the work when the timing is most convenient for you and your business (something a local carrier cares very little about). If you do not have a hardware vendor, look in your phone book under Telephone Equipment & Systems–Installation–Repairing & Service. You can do an online search with these terms, along with your city, to find a vendor. For example, if you're looking for a vendor in the Los Angeles area, open your Web browser and browse your way to the Google home page. Type **telephone equipment systems installation repair service los angles** and click the Google Search button. Any business listed in the results should be able to give you a quote for running wire from the MPOE to anywhere in your office.

If you aren't a thrifty shopper, but like security, it is better to pay a little more and let your local carrier run the inside wiring if that option is available. If the local carrier installs the inside wiring, it probably also offers inside wiring insurance. For a nominal fee per month, this insurance can save you money if a technician has to be dispatched to your office on a trouble issue. If you're generally resistant to having your local carrier dispatch a technician to your site because the fee the local carrier charges is about the same hourly rate as a Ferrari mechanic, remember that having inside wiring insurance means you shouldn't care — the site visit is free.

Adding a new phone line to your long-distance carrier

At some time during the call to add your new phone line, you find out your new phone number(s). Write down the number and then immediately give it to your long-distance carrier. Although it seems logical that your local carrier will tell your long-distance carrier about your new phone line, that is not always what happens.

Despite the fact that all carriers make their money by facilitating communication, you should never assume that your local carrier and long-distance carrier have great communication with each other. Many scenarios could prevent your long-distance carrier from knowing about a new phone line. If the number assigned to your company isn't in sequence with your other lines, the long-distance carrier may not notice the addition. Or perhaps the new line is for a new office down the road from your main office. If the new office receives its own bill, the person doing data entry for your local carrier may not make the connection. Or maybe Mercury is in retrograde and the information just isn't flowing like it should. Blame astrological events. Blame the ironic forces that make it possible for these companies to have vast communications networks and still function in silos. Whoever or whatever you blame, just make the call to your long-distance carier and tie up the loose ends.

Dealing with a screw-up between the local and long-distance carriers

If your long-distance carrier doesn't receive the information to add phone lines to your account, you will suffer one of the following fates. Both fates are equally painful, but easily fixed (they both — shock — involve calling your long-distance carrier):

> **Fate number 1 — outrageous billing:** If a long-distance carrier receives calls from a phone number it doesn't recognize, the carrier charges the highest rate possible. You read that right. The dime-a-minute rate it took you two months to negotiate isn't applied to that new phone line. Instead, your local carrier can bill you at the highest rate on record for the long-distance carrier you selected.
>
> **Symptoms:** When you receive your invoice, you see that you are paying 75 cents per minute with a $2.50 connect fee.
>
> **Solution:** Call your long-distance carrier and have it place that phone line on your account. The problem is that the long-distance carrier may not be able to make the rate retroactive and credit you for the difference on the calls that have already been billed. You may be able to negotiate for a credit, and if the contested amount of the bill is $5 or less, the carrier

may be agreeable. If you are asking for a rerate credit of $600 and you bill out an average of $10 per month, you can expect some resistance. You may be able to get some relief by escalating the issue to the supervisor at your long-distance carrier, but be prepared for the long-distance carrier to stick to its guns.

Fate number 2 — fraud block: Long-distance cariers have either an *open* network or a *closed* network. If the network is open, anyone can complete calls on the network, even if the caller doesn't have a service contract with the carrier. Without a contract, individuals are billed at the highest rate possible, but the network will complete any call it receives. A closed network is the opposite; it does not accept calls from just anyone.

Symptoms: The only way you can complete a call on a closed network is if your phone number has been placed into the long-distance company's system prior to making your first call. If you attempt to place a call on a closed network, your call is rejected and your phone number is logged into the system as a number that is restricted. Because the closed network doesn't have a billing reference or account set up for you, the carrier is concerned that you are going to run up a phone bill of thousands, or hundreds of thousands of dollars, and then disappear. These preemptive blocks, called *fraud blocks,* prevent this problem. Open networks also have triggers for fraud blocks. You may be able to dial phone numbers in the U.S. for months, never setting up a formal account, without having any problems completing calls. Whenyou make two calls to any international destination, however, your second call is blocked, and suddenly you won't even be able to make domestic calls.

Solution: To remove a fraud block, call your long-distance carrier and make sure that it has an account for the phone number that's experiencing the problem. After the account is updated to include the new number, tell the service agent to remove the fraud block.

Changing Your Long-Distance Carrier if You Have Regular Phone Lines

When you have established the phone service for your home or office, the most common change is replacing the long-distance carrier. The long-distance per-minute rate you signed up for years ago is probably higher than you can get today, and so moving your service every few years can definitely save money.

For the sake of simplicity, I group all nondedicated phone lines into the category of *regular phone lines* because the process to change long-distance carriers is the same on all of them. Any phone line that doesn't arrive into your office on one big cable that your hardware splits out into some multiple of 24 phone lines is, by this definition, a regular phone line. Regular phone lines have lots of different names, according to your local carrier:

- **Residential lines:** This is more of a billing designation, but is a basic phone line that terminates to a residence.
- **POTS lines:** I am not lying to you when I say that *POTS* means plain old telephone service. It is a generic term for a regular phone line.
- **1MB:** This is a 1 Measured Business line. This term is virtually synonomous with POTS, except that POTS refers to any regular line and 1MB refers to lines that specifically terminate at a business.
- **Centrex/Centranet:**These are special business lines with a wealth of features available on them.
- **Switched lines:** This is a generic term, like POTS, that referrs to any phone line that isn't running over a dedicated circuit.

No matter what you call them, the industry has a standard way of changing the long-distance carrier on regular phone lines.

Beginning the process

The first step in ordering long-distance service is filling out an application for service with the carrier. The application includes a section that acts as a *Letter of Authorization (LOA).* The LOA is the legalese that allows your long-distance carrier to send an order to your local carrier to point all of your long-distance calls to its long-distance network. In most parts of the U.S., the moment an authorized person signs his or her name to the application or is recorded on tape as agreeing to the service, the order is set and begins running through the system.

But if you're a residential customer in California or Georgia, you must, by law, complete a *third-party verification (TPV)* before your order can be processed. TPV is an automated system that verifies new long-distance orders. Customers must call the toll-free number, access the verification system, and give their name, address, the carrier you want to change to, and phone number.

Every carrier that provides residential service in California and Georgia may have its own twist on TPV services, so you may be asked additional questions. However, the general format is the same. As you are reciting your information, the system tapes everything you say and sends the verification to your new carrier so your order can be processed. In these states, this is a process that every carrier must follow, regardless of their size or track record.

Understanding the timeline

When all the paperwork is complete and you are through the legal hurdles, your new long-distance carrier has to submit the order to your local carrier and then wait for a response. Your local carrier is the company that controls all calls on regular phone lines, routing them to your assigned long-distance carrier, so making sure the local carrier is in the loop is essential. You have to wait about three to five business days for the local carrier to process or reject your order. For information on rejection, see "Explaining your rejections," later in this chapter.

When the order is accepted, your local carrier assigns your new long-distance carrier to your phone lines. From this point forward, your long-distance calls are routed using the long-distance company's network.

Explaining your rejections

If your order for long distance is rejected by your local carrier, all is not lost. There is an industry standard for this process and some general guidelines to prevent orders from dropping off the face of the earth.

An order can be rejected for many reasons, and the industry has created a series of four-digit codes called *Transaction Code Status Indicators* or *TCSI codes,* to help relay the status of your order, including information about whether it is confirmed (which means it was completed), or rejected (and why it was rejected). Every local carrier has its own spin on the TCSI codes, but some of the numbers are standard in the industry. Table 7-1 shows a small section of TCSI codes and gives you a sample of what they look like. The first two digits give you a general idea of the status of the order, and the last two digits give you the specifics.

Table 7-1 TCSI Codes

TCSI Code	*Description*
20-04	Confirmed
20-05	Confirmed Moving
21-66	PIC Restricted (PIC Freeze)
22-15	Disconnected — Working Telephone Number Only
23-28	All Customer Information Changed

Table 7-1 represents an extremely small sampling of the codes that can appear on a migration request for a new long-distance carrier. As you can see, the codes that start with 20 are completed; codes that start with 21 are simple rejections. Your order could be delayed or rejected for any of several thousand reasons. If your carrier already sees your phone line as disconnected (TCSI code 22-15) the carrier won't change the long-distance carrier. Similarly, if your company just merged with another company, and your billing address and company name changed (TCSI code 23-28), the order is rejected.

When your long-distance carrier receives the TCSI codes, you may get a call to confirm that the number is active, that there haven't been any changes placed on the order, or to ask you to remove the PIC freeze from your line. Any rejections that occur delay the migration to your new carrier at least three to five more business days.

Completing the move without rejections and other hassles

The easiest way to change long-distance providers is to call your local carrier and request the new carrier. The process is called *self-PICing your line* because you're personally calling in to have your local carrier change your PIC (*primary interexchange carrier* or long-distance carrier).

The PIC, and sometimes called the *CIC,* for *carrier identification code*, is the four-digit number that your local carrier uses to identify your long-distance network. Table 7-2 gives you a glimpse at some PIC codes and the carriers they represent.

Table 7-2 PIC Codes

PIC Code	*Carrier*
0222	MCI WorldCom
0288	AT&T Communications
0333	Sprint
0432	Qwest Communications
0444	Global Crossing Telecommunications, Inc.

Call your long-distance carrier and ask for its PIC code before you self-PIC. Every long-distance carrier has at least one PIC code, and you need to know it, especially if you are changing to a lesser-known carrier. This isn't because changing your long-distance service to a lesser-known carrier is particularly difficult for your local carrier than changing it to Sprint, AT&T, or MCI. The issue is that the smaller companies evolve over time and are more likely to be bought, sold, merged with other companies, and experience name changes. Keeping information up-to-date isn't always a local carrier's first priority.

If you call Verizon to self-PIC to Broadwing with a PIC code of 0948, for example, your local provider may have the company name listed by its former name, IXC Communications. Or if your local carrier was Bell South, Bell South may have Broadwing (also known by its previous incarnation, IXC Communications) listed as Switched Services. All three entities have the same PIC code, but the list used by the local carriers to cross-reference the code with the name may be years out of date, hundreds of companies long, and sorted by company name, not PIC code. Knowing the right code prevents most local carriers from searching for the PIC code on your behalf, and reduces the possibility of errors. You can also give the local carrier options on names so that it can validate the four-digit PIC. In the end, the incarnation of the name they have listed doesn't matter, as long as the PIC code is correct.

Some local carriers ask for a five-digit rather than a four-digit PIC code. If this question is posed to you, don't worry. Simply add 1 to the PIC code your carrier gave. For example, if you're switching to AT&T, instead of telling your local carrier the PIC of 0288, simply say that it's 10288.

Keeping bills within reason when you self-PIC

You need to be aware of a couple of potential problems so that you avoid sky-high phone bills when you self-PIC:

- **If you give your local carrier the wrong information, you risk being blocked for long-distance service or being billed at very high rates.** If you make a mistake, the erroneous carrier you request will have no record of the order and will charge you as much as it can. And because you made the change yourself (you decided to self-PIC; the company didn't slam you), you don't have much leverage to ask for a credit or any leniency. The worst part about this potential problem is that you will probably not see your first invoice from the new carrier for about a month, by which time the damages could be substantial.
- **Before you self-PIC to your new long-distance carrier, confirm that the long-distance company's network is set up to receive calls from you.** The exact same problems could crop up if you are self-PICing to your long-distance carrier or if somehow you are sent to the wrong network. The only benefit you have if you are billed at the maximum rates from your carrier is that the long-distance carrier has a vested interest in keeping you happy. If you chat with a representative of the new company, you may be offered a courtesy credit or a rerate.

Confirming your long-distance carrier

After you sign the contract with your new carrier, your contact at the new long-distance carrier will certainly attempt to reassure you that in about seven business days you will be enjoying the new lower rate. That is great, but the timeline may not be met for one of a variety of reasons. Because you don't work for a carrier, you will not get to see the TCSI codes to check on whether the order was rejected or completed. The good news is that you can test your phone lines to confirm your long-distance carrier any time of the day or night.

A *700 test* is an industry term that refers to dialing 1-700-555-4141 on your touch-tone phone to determine long-distance carrier. The 700 test number is a free call that directs you to a recording set up by your long-distance carrier that simply says something like, "Thank you for using AT&T." The test is 99 percent accurate.

The 700 test only validates the long-distance carrier for the individual phone line being dialed on. If you have 70 phone lines you want to check, you have to manually seize each phone line and dial the 700 test number. There isn't a way to check a group of phone lines with this test. If you had an open order to add voicemail to your main phone line, or to change the contact name on the account, or any activity on any of your phone lines, the open order would have blocked your line from being changed. This scenario is much more common than you might expect, especially for large customers, to find a handful of lines that didn't migrate to your new carrier for one reason or another.

The 700 test only validates the PIC your local carrier has listed for your interLATA long-distance traffic. Some local carriers offer a 700 test for your intraLATA PIC of 1-700-(your area code)-4141, but generally, you can only confirm the long-distance carrier for your interLATA calling. If you aren't really sure what intraLATA calls are versus interLATA calls, check out Chapter 3 for the specifics.

Sometimes 700 tests do fail, but for the majority of people they work perfectly. Some *competing local exchange carriers* (CLECs) like PAETEC or Mpower may not provide the feature, but may instead send you to an automated voice that recites the phone number you are dialing from. It is uncommon, but sometimes carriers that provide 700 tests may have conflicting information in their systems; the 700 test may not be routed anywhere. When in doubt, call your local carrier. Your rep should be able to validate everything you want to know.

Cancelling service with your old carrier

You have confirmed your lines are set up with your new carrier, and the 700 test says that you are on the new carrier, so you are done, right? Well, no. Now you have to call your old long-distance carrier and cancel your existing service. One section of the long-distance company may receive a report from your local carrier that explains that you have changed carriers, but that doesn't mean that your old carrier has any incentive to stop charging you, unless you call in and cancel the service.

It is unlikely that your long-distance will change from your old carrier to your new carrier at the exact moment that your billing cycle ends. You will receive an invoice from both carriers for the same month you transition to the new carrier. It is important to check this invoice to ensure all of your lines in all of your areas of coverage (intraLATA and/or interLATA) have moved to the new carrier.

You may receive invoices from your old and new carriers for the same month at about the same time. Don't worry; you are not being double-billed. Check the detail sections of each bill to see whether any of the calls overlap. There should be a date when calls stop on your old carrier and start on your new carrier. If you see no obvious sign that the calls stop with the old carrier and begin with the new one at around the same time, call your new carrier for help.

Casual Dialing

The PIC code you give to your local carrier when you self-PIC also has another use in the world of telecom. It acts as the basis of the *dial-around code* you can use to force your call onto the long-distance provider's network.

Dial-around codes are the best gadgets in your telecom toolbox for switched phone lines. They allow you to use a long distance carrier's network even if your local carrier doesn't send your calls there by default. Dial-around codes also provide you with alternatives if your primary carrier is having a network failure. Dial-around code features do come with potential dangers, so read the following section to understand the pros and cons of using them, and then determine who in your office needs to know how (and when) to use them. If information about the existence of dial-around codes falls into the wrong hands, your business could lose a lot of money when the phone bill arrives. But in the right hands, they can save your company a bundle.

Dial-around codes became popular in the 1990s when companies began marketing that customers can simply dial 10-10-321 plus 1, and then the area code and number to make calls at ten cents per minute, anywhere in the continental U.S. As a result, dial-around codes are also known as *10-10* (referred to as ten-ten) *codes* or *trick codes*.

C*asual dialing* allows you to have all of your outgoing calls process by using your normal long-distance carrier by default, except for the calls that you dial with the corresponding 10-10 code.

Understanding how 10-10 codes work

Dial-around codes override the routing with your local carrier. Usually, when you dial a long-distance number, the call is transmitted to your local carrier; the local carrier identifies the long-distance carrier for your phone line, and sends the call off to the long-distance network. The 10-10 code stops that process short. When your local carrier receives a call that begins with a 10-10 code, it understands that to mean "do not pay any attention to the long-distance carrier that is set up on this phone line; simply place the call on the network with the last four digits of this 10-10 code." If you dial 10-10-288 and then a phone number, your local carrier ignores your chosen long-distance service provider and sends the call by using the carrier with PIC 0288 (which happens to be AT&T, in case you were wondering). A comprehensive list of the PIC codes — how they look as dial-around codes can be found at Discount Long Distance Digest, the Internet Journal of the Long Distance Industry (`www.thedigest.com/faq/picodes.html`).

Calls placed with 10-10 codes are billed by your local carrier and itemized in the long-distance section of your phone bill. If you casual dial using several carriers, your phone bill will have sections for each casual dialing carrier, listing some general contact information for the carriers and an itemized list of individual calls made with each carrier in their network.

Benefits of casual dialing

Because casual dialing enables you to have calls routed to the long-distance carrier of your choice, it gives you a lot of freedom. Some of the benefits of using a casual dialing code are:

- **You're never without long-distance service.** If you have a dial tone on your phone, and your local carrier hasn't placed any restrictions or blocks on your line, you can use any carrier with an open network whenever you want. When a large earthquake hit Northridge, California, in 1994, most of the major carriers' networks went down. Survivors of the quake who had long-distance service with MCI or AT&T were out of luck — all outbound calls failed to a fast-busy signal. On the other hand, the Sprint network was still working fine. Individuals who dialed 10-10-333 for Sprint and then dialed 1+ area code+ telephone number were able to contact loved ones without a problem. Of course, those people may have been charged $3 per minute for the call, but at least family members knew that all was well.

  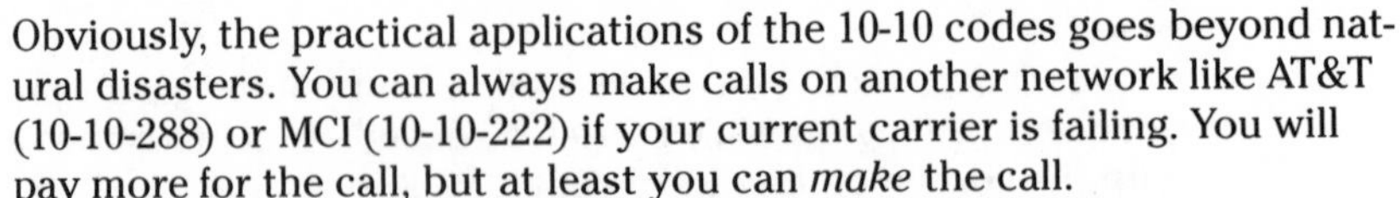

  Obviously, the practical applications of the 10-10 codes goes beyond natural disasters. You can always make calls on another network like AT&T (10-10-288) or MCI (10-10-222) if your current carrier is failing. You will pay more for the call, but at least you can *make* the call.

- **No default taxes per phone line are charged to you by the 10-10 carrier.** Because you're using the service only when you want to, and on only the calls you want, the standard fees assessed for having a phone line assigned to a long-distance carrier don't apply.

- **Your local carrier doesn't charge you to change carriers.** You have the benefit of placing calls by using another carrier, but don't have to wait for your local carrier to change the PIC on your phone line, and you don't have to pay your local carrier for the privilege.
- **You can try it before you buy it.** If you would like to move to another long-distance carrier, but you have reservations about the quality of its network, have the carrier set up your lines for casual dialing before you authorize a full-fledged PIC. (The carrier's technicians open the network to receive and process calls from your phone numbers, but don't send an order to your local carrier to make any changes.) During that time, you can make calls with their 10-10 code to your normal calling areas to ensure that the quality is sufficient. This is a very useful suggestion if your organization makes a lot of international calls. The international platforms that carriers use can vary quite a bit, and you don't want to jump to a new carrier if there is so much static on the lines when you do get through to Asia that it negates any financial savings you might gain.
- **You can set up your phone line on more than one carrier.** If you have a great deal on your domestic calls with Qwest, but its international rates are less than stellar, you can use another carrier for those calls. You need to ensure that your backup carrier is okay being set up for casual dialing only; but assuming that the second carrier doesn't have an issue with it, you can get the best of both worlds. After the carrier tells you the 10-10 code you need to use, and opens up its network to receive your calls, you can instantly begin on the network. All of your domestic long-distance calls are sent as usual with your primary long-distance carrier's network. If you want to make a call overseas, say to Germany, you simply dial 10-10-321 (or whatever the access code is) +011 49 and the rest of the number, and the call completes by using your casual dialing carrier.

Making sure your casual dialing carrier is designated as no-PIC

When you speak to the carrier you want to use for casual dialing, you must confirm that your system will be set up correctly. You don't want an order sent to your local carrier to change the long-distance on all of your phone lines.

The casual dialing partner's system must designate your numbers as *no-PICs*. As long as the carrier you are speaking to provides casual dialing, the agent you speak to should recognize the term no-PIC when you throw it around, and understand that you don't want your local carrier to have any formal relationship with this carrier. Specifying the status of your relationship as no-PIC ensures that your lines are open on the casual dialing carrier's network, but that they don't disturb the configuration with your local carrier.

Avoiding the downside of casual dialing

Casual dialing offers quite a few benefits if you plan it out, research it well, and use it wisely; but casual dialing comes with some challenges as well:

- **You may forget to dial the 10-10 code before you dial the phone number.** The 10-10 code only works if you dial it before every call you want to make on the casual dialing network. If you forget to dial the 10-10 code before the phone number, the call defaults to your standard long-distance carrier.
- **You may forget the 10-10 code and dial the wrong digits.** If the 10-10 code you use is 10-10-345 and instead you dial 10-10-543, your call is sent to another carrier. If the other carrier's network is closed, your call fails. If the other carrier's network is open, the call goes through, but your bill may be much higher for the call than you were expecting.
- **You may have little (if any) customer service access.** If you use a 10-10 code that was advertized on TV or the Web, you won't have access to a large support network. Who do you report trouble to if you dial the 10-10 code and your call fails, or has static, or you heard so much echo that you have to hang up and call back using your regular carrier? Who do you call to dispute your invoice if there is a 3-hour call listed on it from your phone that overlaps 15 calls you made on the same line by using your primary carrier? The support networks for companies that *only* provide casual dialing access, and no other telecom services, is generally not as robust as you will find with a mainstream carrier that also offers the service along with other calling services.

Moving Your Phone Number

LNP stands for *local number portability*, and it is a procedure that enables you to move your phone number from one local carrier to another. The procedure came into being because people hated getting new numbers every time they switched cellphone providers. After enough people got irritated, the U.S. Congress forced the development of the LNP system. Now if you switch wireless providers, you can take your phone number with you to the new one.

In the context of wireless providers, the LNP system seems like small potatoes. Now consider the fact that wireless providers also function as local carriers; you can move your phone number from local carrier to local carrier, which is actually, well, big potatoes.

Here is a real-world scenario that could happen today: You live in Charlotte, North Carolina. You're a busy executive, so on any given day you may be in the office, on the road, or working remotely from home. You don't want your staff or associates to be forced to dial three different phone numbers to reach you, so you decide to *port* your phone line to a competing local exchange carrier (CLEC) that uses VoIP (Voice over Internet Protocol) technology to provide a *find-me-follow-me* service, completing calls dialed to one phone number to multiple destinations. (I discuss porting in the following section.) The increased functionality may motivate you to move your phone line to the new carrier, and you may not even have to upgrade to VoIP phones at your office. If you are interested in VoIP, please read more about it in Chapter 15.

Understanding the Porting Process

Cellphone and wireless providers have been porting phone numbers for years. In most cases, the process has become an efficient transaction that may take as little as a few hours. However, it is by no means a seamless process, and not without its limitations. If you want to *port* your phone number from one local carrier to another, you need to bear these facts in mind and adjust your expectations accordingly. Porting a phone number is not always seamless, and you may have days, or weeks, during which your phone number is completely inoperable. There are systems in place to reduce the downtime, but downtime is always a possibility.

Porting your phone number is a lot like ordering new long-distance phone service. You fill out an order form, agree to the terms and conditions, and then sign a *Letter of Authoriztion* (LOA) to enable the new local carrier to begin the porting process on your phone line. From that point forward, nothing is standard. Some of the issues you may face are:

- **Your phone number is not actually portable.** If the local carrier to which you request to port your number has not been authorized to be a CLEC in your state, it can't port the number onto its network. Or, if the local carrier to which you request the number be ported doesn't have a porting relationship with your existing local carrier, it can't port the number. Most carriers ask you to validate that your number is portable before you begin the process so that you don't find out this information a week into the process.
- **The time frame for completing the port varies.** If the local carrier you are moving to has forged a strong relationship with your old carrier, you may have your number moved over in two weeks, with negligible downtime. If your current local carrier is obstinate and doesn't play well with others, the process may take six weeks, and you may not have phone access for three days of that time.
- **You may experience an unfortunate name mismatch.** If you submit a request to port a number, and you list the name on the account as Bob Smith, but your local carrier has the number listed under your wife's name Mary Mahoney, the porting request is rejected. You may need to resubmit the port request under the name Mary Mahoney, or, if the

name on your local carrier invoice is Bob Smith, send in a copy of the bill to dispute the rejection.

- **The port request may be rejected because of a pending order on your phone lines.** If you ordered voicemail from your local carrier months ago, and the order was never completed for one reason or another, any port request is automatically rejected. Voicemail isn't the only source for this glitch; any open orders on your local service to add, change, or disconnect a service, prevent your lines from porting. If the service is unnecessary, or simply got lodged in the order system with your local carrier, a quick call to cancel or accept the new feature is all it generally takes to cross this hurdle.
- **The port request is rejected because you have a Centrex/Centranet system.** If you have any packaged business features from your local carrier, there may be limitations on what can be ported. Centrex and Centranet products generally have a requirement that three lines must remain active on your account for you to be offered the packaged business services. You may be required to port all the lines in the Centrex or Centranet group, or you may be able to port some of them — as long as you leave three lines active on your old local carrier. You need to discuss the possibilities with your current and future local carrier.
- **The port request may be rejected because you have nonportable services on your line.** DSL is the main culprit in this group. If you port a phone line that has DSL, either the port is rejected, or the port completes but you lose your DSL service. You will have to call your local carrier to separate your phone line from your DSL. Local carriers commonly allocate a small section of their bandwidth available on the cable that supplies DSL to be used for regular phone lines. Because your phone line and DSL cable are connected on the same piece of copper wire, removing one of them potentially cancels the other service. Splitting the services onto individual copper wires prevents the loss of Internet service when your voice line ports away. If you have a distinctive ring, or some other feature that may not be available with your new carrier, you will have to decide to either lose the service or abandon the move.

Chapter 8

Ordering Dedicated Service

In This Chapter

- Understanding your configuration information
- Ordering ISDN and SS7
- Knowing your local loop pricing and options
- Receiving the Firm Order Commitment (FOC) document
- Using the Carrier Facilities Assignment (CFA) document
- Understanding the Design Layout Report (DLR) document
- Selecting a collocation provider
- Creating a technical cut sheet

You can use dedicated circuits for many things, including to aggregate standard voice lines and to provide the access for your Internet connection. Similarly, dedicated circuits can terminate to your local carrier (for a local dedicated circuit) or to your long-distance carrier (in the case of a long-distance dedicated circuit). The process of ordering a dedicated circuit is generally complex, but you can offload much of the process to your carrier and hardware vendor. Even then, the process can be frustrating if you don't understand the steps involved.

Every carrier has its own timelines for ordering dedicated circuits, and delays aren't uncommon. Simple problems may slow you down by 24 hours, and large problems can set you back several months. This chapter covers everything you need to know when you order a dedicated T-1 circuit, and helps you know what to expect so you won't be surprised when your carrier quotes you 30 to 45 days to install a small circuit. You may need to order a circuit larger than a T-1 to meet your telecom needs, but in the end it all boils down to the T-1 (DS-1) level circuit and the 24 individual voice channels (DS-0s) that allow you to have a conversation. If you aren't sure a dedicated circuit is for you, brush up on the financial justifications for a dedicated circuit in Chapter 2.

Ordering the Circuit's Configuration

Before you can begin to order your circuit, you need to know your circuit's configuration. A dedicated circuit must be built the same on your phone system that it is on your carrier's hardware in order to function properly. If your hardware is set up to speak B8ZS/ESF line coding and your carrier is set up for AMI/D4, you won't be able to communicate. This is similar to a person who only speaks Swahili trying to talk to a person who only understands Russian. Until they are conversing in the same language, nobody will understand anything. In the telecom world, you have to worry about the protocol used to set up and tear down calls as well as the location of these messages to allow your hardware to communicate effectively with the network.

To avoid having to understand this entire section on configuration, ask your hardware vendor to fill out the technical part of your carrier's order form. It's better to have someone who knows your hardware to fill out the form, than to try to figure out the techie stuff and do it yourself. If you are replacing existing circuits, ask your current carrier for your circuit's configuration. If you're coming to the end of your term with the carrier and you've had consistent problems for the past six months, the carrier might be suspicious that you're thinking of switching services, but the carrier should still give you the information. Use the following sections to familiarize yourself with the information your new carrier will need. After you have a general grasp of what the new carrier needs, go through the technical worksheet and fill out what you can. When you find a section you're unclear about, ask the carrier for clarification, and then pull the information from your hardware or your current carrier to fill out the section.

Every bit of the configuration is crucial when ordering a dedicated circuit. The greatest challenge is that there aren't standard package configurations that everyone uses. The way your phone system is set up is uniquely tailored to meet your company's needs and limitations, the features of your phone system, and the options available from your carrier.

Speak with your hardware vendor whenever you think about changing your carrier or your phone system. The hardware vendor can be a tremendous help in determining the costs, timelines, and limitations of your system.

Time division multiplexing and clock source

Time division multiplexing, or *TDM,* is a technology that enables you to take a T-1 circuit that passes 1.54 Mbps and parse it out into 24 individual channels of 56 or 64 kbps. TDM does this in a methodical and controlled manner so that your carrier can decode and recombine data, using the same methodical and controlled manner. This technology is quite complex, and makes up the protocol for the vast majority of dedicated circuits used for voice calls in the world. Fortunately, you only need to know the basics to use it effectively.

TDM is based on sampling small sections of phone calls very often. The easiest analogy to imagine is that there are 12 sets of people who want to have conversations on the same phone line. In order to make sure everyone receives equal time, a stopwatch dictates when each pair is given its turn to speak. Because 12 pairs of people are speaking, each group can speak for only five seconds every minute. As the hand of the stopwatch passes from 12 to 1, the first couple has five seconds to speak. When the second hand passes from 1 to 2, the second couple has five seconds, and so on, down the line. After you speak for 5 seconds, your phone goes dead until everyone else has had their time, and then 55 seconds later, you can speak for another 5 seconds. Figure 8-1 shows the five seconds you have to speak, before you have to wait for the clock to return to the 12, at which point you get your time again.

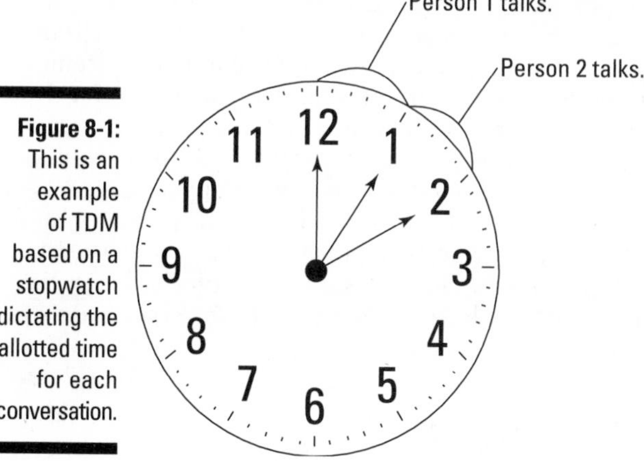

Figure 8-1: This is an example of TDM based on a stopwatch dictating the allotted time for each conversation.

Asking people to speak for only five seconds per minute is impractical, so the world of telecom speeds the process up. Imagine the stopwatch moving faster. Instead of the second hand passing around once a minute, imagine the stopwatch sped up to complete an entire cycle once a second. Now you have to wait only a fraction of a second before you can speak again. The problem is that you still only have 1⁄12 of a second to say anything, so your conversation would be very choppy and generally incomprehensible. The beauty of TDM is that your conversation isn't sampled only one time a second, but 3,000 times per second, so you have to wait only a few milliseconds before it's your time to speak. Because what you're saying is being sampled and transmitted so often, you don't even know that 23 other conversations are going on around you on the same T-1 circuit. However, both phone systems still know who gets the correct timeslots; your hardware packages your call in the correct timeslot and sends the call down the line to your carrier, who uses the same system to unpackage and combine your call on the other end. This system allows everything to flow down the line without information being mixed up or lost.

With a TDM system, your hardware has a clock, and so does your carrier. The two clocks are perfectly synchronized (like two dancers) so that they can both know when to begin and end a call's timeslot. The system functions like a synchronized waltz between the two pieces of hardware, and just as with dancing, someone has to lead. Your carrier should always be configured to lead, — to be the primary clock source — that dictates what the exact time is.

If you have multiple circuits from different carriers, you must confirm that your hardware is capable of assigning the primary clock source individually, on a per-circuit basis. Maybe your hardware is designed to handle multiple T-1 lines, but if it can't identify more than one primary clock source for the group of circuits, you face some pitfalls. If all of your circuits are coming from the same carrier, this software configuration is fine, because all the circuits will probably use the same clock source from the group. If you have circuits from four carriers, each circuit needs to use the clock source from its own carrier. If your card handles four circuits, yet pulls clocking from one circuit and applies it to the remaining three, you will experience cumulative frame slips and errors on the other circuits. The frame slips represent your hardware failing to synch up with your carrier; this situation may cause echo or static on your calls, or it may have no effect until the circuit is so far behind the carrier clock source that the whole system crashes. After you restart your hardware, everything will work just fine — for a while, until your system falls behind and everything crashes again. The bottom line is that your hardware must be able to handle a primary clocking source for every carrier. If it doesn't, and you want to use TDM, you'll need to upgrade your hardware.

Line coding and framing

The line coding and framing of a T-1 define the *overhead* and *useable bandwidth* available for each of the 24 channels. The overhead of a channel works with the clocking and is responsible for the maintenance and housekeeping on a call, including the call setup and tear down, ringing, and those fun recordings that you get sometimes that tell you the number you have dialed is disconnected or is no longer in service. The useable bandwidth is everything left on the dedicated channel that isn't overhead and is used to transmit the speech in the conversation. There are only two options when it comes to line coding and framing. You can have

- **B8ZS/ESF** (which stands for Binary 8 Zero Suppression/Extended Super Frame — say that five times fast). Traditionally, data transmissions are passed across B8ZS/ESF circuits. These circuits have more useable bandwidth and less overhead.

- **AMI/SF** (which stands for Alternative Mark Inversion/Super Frame, sometimes also called *D4/SF*). Voice transmissions are traditionally passed across AMI/D4 circuits. These circuits have less useable bandwidth and more overhead.

As you can see in Figure 8-2, both T-1s have 24 channels, with the top shaded portion identifying the overhead for the circuits.

The increased overhead on the AMI circuit allows for more of the bandwidth to synchronize voice transmissions. Circuits that are used to transmit data need less overhead, because any problem in transmission can be overcome by resending the data.

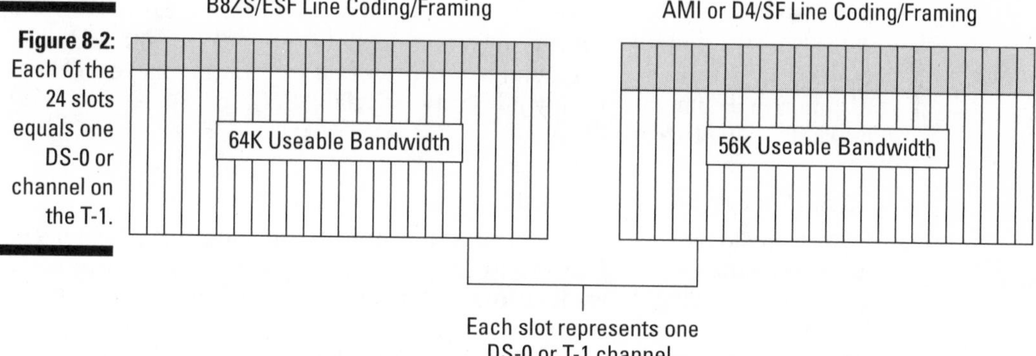

Figure 8-2: Each of the 24 slots equals one DS-0 or channel on the T-1.

Ensure that you order the correct line coding and framing on your circuits. If you are ordering a long-distance T-1 and AMI/SF is delivered instead of B8ZS/ESF, correcting the mistake can easily take about two weeks. This is because the local carrier that delivers the circuit has to physically show up at your office to reconfigure their hardware. As long as the line coding and framing are correct, you can generally change the lower-level protocols, such as the outpulse signal and start on the fly when you activate your circuit. This is generally not the procedure for most carriers, but, as long as you get a warm-hearted installation technician on the line, you'll be well taken care of.

Outpulse signal and start

The *outpulse signal* and *start* are two protocols working together to identify how calls are initiated and terminated. The outpulse signal is used to set up and tear down calls. The start dictates the required sequence of events. For example, the start determines whether your carrier needs to wait to send users a dial tone, or whether the carrier needs to send you a signal first to acknowledge that users want to hear a dial tone. When you want to make a call from your home, you pick up your phone, at which point a signal is sent to your local carrier to send you a dial tone. A few milliseconds later, you hear a dial tone, press seven or ten digits, and your call is processed.

In the world of dedicated circuits, you have options about how you request a dial tone, and whether the overhead consists of a one-way or a two-way conversation. There are only three options for line coding and framing. See the following sections for more information.

Loopstart

Loopstart is a protocol whereby the carrier is always waiting for you to make a call. Loopstart is essentially a one-way conversation, but it's an effective and simple protocol.

Groundstart

Groundstart is another one-way conversation, but from the opposite end of the circuit. In this case, you tell the network when you want to make a call. It's more of a command from your hardware telling the carrier "Hey, I'm making a call, process it!"

E&M Wink or E&M Immediate

E&M is an old telecom protocol that stands for *e*ar and *m*outh. It's the only bidirectional protocol of the group and has its niche in the telecom world. Instead of receiving orders like groundstart or waiting for calls like loopstart, the E&M protocol is like a constant conversation between your hardware and the hardware of your carrier. It looks something like this:

> **Your hardware:** "I'm going to make a call; give me dial tone."
>
> **Carrier:** "Okay, here is a dial tone."
>
> **Your hardware:** "I'm sending the phone number; do you have it?"
>
> **Carrier:** "I got the phone number and I am processing it."
>
> **Carrier:** "The call reached the final office; I am sending you a ring tone now."
>
> **Your hardware:** "Okay, I got the ring tone."
>
> **Carrier:** "Someone answered the call; I will stop sending the ring."
>
> **Your hardware:** "Okay, the ring has stopped."
>
> **Carrier:** "I am sending you the conversation now . . ."

E&M protocol comes in two flavors, E&M Wink and E&M Immediate.

- **E&M Wink:** You have the full conversation with E&M Wink. This might seem chatty, but if you have toll-free numbers that are riding on your circuit, you might need E&M Wink in order to use all the information and features sent in the call's overhead.

- **E&M Immediate:** If you aren't using toll-free numbers, E&M Immediate might be a faster and more efficient protocol.

If your circuit will be processing calls for toll-free numbers that require DNIS or ANI delivery, you must use E&M Wink. Groundstart, Loopstart, and E&M Immediate don't provide the bidirectional protocol required to use these dedicated toll-free features. Other *out-of-band* signaling protocols provide DNIS and ANI delivery, but they have to be offered as options on your hardware. Skip ahead to the section called "Feature Group D" if you are interested in these other protocols. If you aren't sure what DNIS or ANI delivery is, Chapter 5 gives you all the info you need.

Introducing trunk groups

Trunk groups are the contiguous channels within your dedicated circuits that are partitioned off from the rest of your T-1. The trunk groups can split a single T-1 into smaller sections, or they can combine several T-1s into one large group. If you combine three circuits into one trunk group, you can then receive calls on your toll-free number on all 72 channels just as you would with one large circuit. If each T-1 has its own trunk group, your toll-free calls can come in on only 24 channels; otherwise, you have to set up overflow to the other trunk groups, which might cost more.

Trunk groups only affect how your carrier sends calls to your circuit. If your carrier splits your T-1 into 4 trunk groups of 6 channels each, your phone system can still dial out on all 24 lines, in any order you want, without any concerns. Because your phone system determines the channel you dial out on, it isn't bound by trunking parameters set up by your carrier. It simply seizes whatever channel it wants and dials.

Because some trunk groups are large and cover multiple T-1s, there are standard ways you receive the calls within a trunk group. Your options are as follows:

- **Least Idle:** In this case, the channel that has been active the longest receives the next call. Call centers typically use this feature because their fastest employees get more of the calls. Generally, this option can overload a single port on your hardware and might result in failures, so I don't recommend it.
- **Most Idle:** This is a better scenario; the system sends the call to the phone line that has been idle for the longest time. This type of routing distributes calls evenly across all your channels, without burning out one specific channel.
- **Ascending:** Calls are sent into your system in sequential order, starting at channel 1, and going up to channel 24 (or the end of the trunk group). The first channel on your circuit will probably always be active, because

this option continues to backfill the lower channels when they become available.

- **Descending:** The opposite of ascending, calls are sent to your circuit in sequential order, but this time starting with channel 24 (or the top end of your trunk group) and on down.

Try setting up your inbound calls by using descending and configuring your phone system to use ascending for outbound calls. In this scenario, your inbound calls are working down from channel 24 and your outbound calls are hunting up from channel 1. This reduces your chances of *glare*, which is an undesirable condition that occurs when an inbound call and an outbound call attempt to seize the same channel.

Evaluating Out-of-Band Signaling

All the protocols mentioned previously are classified as *in-band* signaling, because the overhead used to manage the call is located within the same timeslot as the voice portion of the call. Refer to Figure 8-2 to see how each channel has a section of usable bandwidth and a section devoted to the overhead and maintenance of the call.

Out-of-band signaling takes the overhead of each call, and, instead of using space on the same channel as the voice portion of the call, it aggregates the overhead on a single channel that might not be attached to your circuit. Out-of-band protocols allow your phone system to receive additional information on each call, or they may simply shorten the amount of time it takes to connect a call. The information you require from your telecom service, the limitations of your existing hardware, and the amount of money you can spend for the features all define the best protocol for your company.

You have three basic options for out-of-band signaling in the TDM world, each with its own strengths and weaknesses. Review them before you chat with your hardware vendor to find out what it takes to integrate them in your phone system.

Integrated Services Digital Network: ISDN

The *Integrated Services Digital Network* protocol, commonly referred to as *ISDN,* is the most common out-of-band signaling option used at the T-1 level. The protocol is standard on many new multiplexers and MUXing cards (such as those made by Dialogic) that can be installed in a PC environment to perform the functions of a channel bank. The ISDN protocol provides a wealth of information available on inbound calls, as well as detailed information for your carrier on outbound calls. DNIS, ANI delivery, and ANI Infodigits are standard features on most ISDN circuits. The reason that ISDN can provide all this additional information on each call is because it uses an entire channel for overhead. Figure 8-3 shows you what the overhead looks like on a T-1 with 24 channels, one of which is reserved for overhead and identified by shading in the figure.

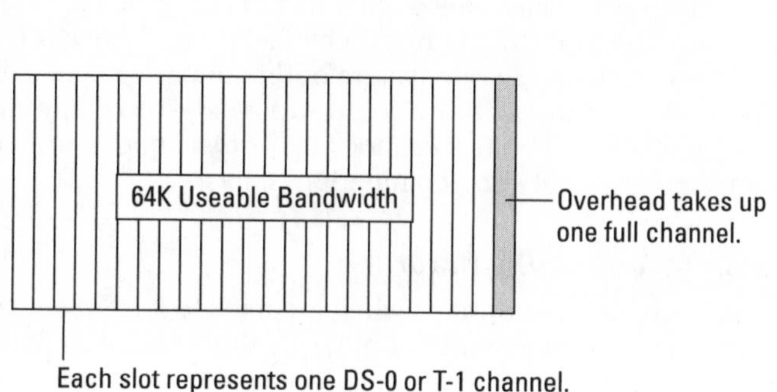
64K Useable Bandwidth
Overhead takes up one full channel.
Each slot represents one DS-0 or T-1 channel.

Figure 8-3: Standard configuration for ISDN, with 23 bearer channels and 1 data channel, which handles the overhead.

REMEMBER

All ISDN circuits should be provisioned as B8ZS/ESF at the line coding and framing level, to allow you 64 kbps useable bandwidth for each channel. Most carriers require you to order the circuit as B8ZS/ESF. It might be technically possible to use the ISDN protocol with AMI/SF circuits, but it just isn't done.

Understanding B and D channels

ISDN circuits consist of channels that transmit the voice portion of the call, called *bearer (or B) channels,* and channels that just handle the overhead, called *data (or D) channels.* In Figure 8-3, the 23 channels that are white are the B channels; they transmit only the voice portion of calls. The shaded channel is the D channel; it transmits the overhead for those 23 B channels. This is a standard configuration for ISDN, with one D channel on the 24th timeslot of every T-1 in the system. Sometimes people like to mix things up and place the D channel on the first timeslot rather than the last one, but it's generally on one end of the circuit.

REMEMBER

There is a lot of space on the D channel — actually more than enough space to handle the 23 calls on a single T-1. To use the D channel's space more efficiently, you can use a single D channel to service multiple T-1s. Some companies hate losing an entire channel just for the sake of having a more advanced protocol. If you can use that one channel for more than one circuit, the loss seems somehow more palatable. You need to remember a few things if you use one D channel for multiple T-1s:

- **All the T-1s must be from the same carrier.** You can't bring in a pair of T-1s from Sprint and another pair from Qwest and expect one D channel to handle them both.
- **You need to confirm that your hardware and your carrier's hardware can handle the configuration.** Talk to your vendor before you make big plans.
- **You can experience problems during peak calling times.** Just because the hardware says you can use one D channel for up to seven T-1s doesn't mean that doing so gives your system optimum performance. When you push the limits of your hardware and try to handle too many circuits with one D channel, your channels can lock up or disconnect, or you can experience other quality and completion issues.

Choosing your favorite ISDN flavor

All carriers offer a handful of options when it comes to ISDN. Some call them protocols, and some call them switch types. The bottom line is that you have to set your hardware to speak the same ISDN language as your carrier. Some of your domestic ISDN choices are:

- **NI1:** National ISDN 1 (Bellcore standard)
- **NI2:** National ISDN 2 (Bellcore standard)
- **5ESS:** AT&T standard
- **DMS100:** Northern Telecom standard

Your carrier might not provide every ISDN protocol listed here. If there is a nuance that makes NI2 perfect for your business, be sure to confirm that your carrier can provide it. Some carriers only offer NI1, and they won't pay to upgrade all of their switches when they are really waiting for NI3 to arrive. You really need to conform to the preferred protocol of your carrier with ISDN if possible, or choose a different carrier. The problems you encounter if your carrier is using a less-preferable protocol may be much more troubling than if you simply used your second- or third-choice hardware. The ten-millisecond reduction in call setup or tear down time isn't worth the increased percentage of timing related call failures.

Using ISDN for telemarketing

You will want to change to ISDN protocol if your primary business is telemarketing. Recent federal regulations require telemarketing businesses to present a valid Caller ID number when placing telemarketing calls. The ISDN protocol enables you to select a phone number to use as your company's Caller ID. You can change the Caller ID number based on the calling campaign, and can even display a toll-free number.

A note if you already have ISDN service

ISDN was initially released, not in the full T-1 variety called PRI (Primary Rate Interface), which is discussed previously, but as a smaller service called Basic Rate Interface (BRI). The service was developed and released to handle the massive demand for desktop videoconferencing features. The BRI service was much smaller than the full T-1 version and consisted of one D channel and two B channels. The service was initially problematic, because the available hardware was not very compatible and installation was troublesome. The key selling point of the BRI was a feature called *bonding,* which allowed users to combine both B channels into a single 128KB path, instead of existing as two individual channels with 64KB of bandwidth. Users could even bond together several BRI circuits to create a 256kbps connection as well. Unfortunately, not many people wanted to videoconference, and as DSL and other technologies were released, BRI lines became less popular. If you have a BRI line, you need to know that most long-distance carriers don't support bonding; AT&T, MCI, and possibly Sprint still offer this support, but who knows for how long? If you have two BRIs and use your 256 kbps connection to transmit CD-quality music from a microphone in Hawaii to a recording studio in New York, that is great, but don't change your long-distance carrier unless you know it also supports bonding.

Currently, you can input only numbers, not letters, into the Caller ID field. Federal law states that you must also list the company name, if your carrier supports the service. The only problem is that no long-distance carrier can support alpha characters (letters) in the Caller ID field. Anything the long-distance carrier receives that isn't a number is stripped when the call is sent to the destination.

ISDN users are frequently charged monthly fees based on the number of D channels used. Every carrier offers different pricing, but watch for hidden costs when using ISDN. Many carriers charge an installation fee and a monthly rate for D channel cards. Ostensibly, the monthly charge is used to recover your carrier's cost to purchase D channel cards for its network, but it seems more like a profit center. To give you an idea of what kinds of charges you might see, your company can conceivably be charged $250 for installation and $200 per month per D channel. If you have 28 T-1s, each with their own D channels, you are looking at an installation fee of $7,000 and a monthly recurring charge of $5,600. Um, did someone say ouch? Before you order your ISDN circuits, check with your carrier about additional fees.

Signaling System Seven: SS7

The *Signaling System Seven,* or *SS7,* protocol is the epitome of out-of-band signaling. The SS7 configuration is even more removed than ISDN, because the SS7 link that handles the overhead is on a completely separate and noncontiguous circuit. Figure 8-4 shows how the SS7 link is positioned in comparison to the circuit whose calls it manages.

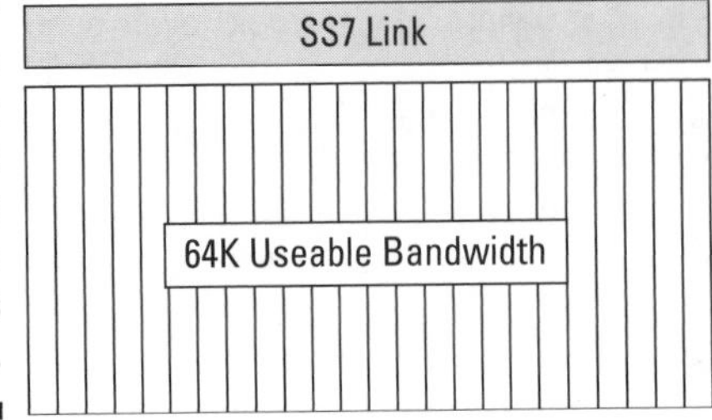

Figure 8-4: T1 with 24 voice channels and an autonomous SS7 link.

The main reason businesses use the SS7 protocol is because they need a very quick connect time for all of their calls. In fact, telecom carriers and large telecom customers often use SS7. If your business experiences an unacceptable level of latency on domestic calls, and you spend more than $50,000 on your long-distance every month, it's time to look into SS7.

Understanding how SS7 functions

The SS7 protocol is like having a scout when you are on safari. The protocol reads the phone number you have dialed and runs out ahead of the body of the call, checking the possible routes to get you to your destination. If it discovers an outage in Dallas, Texas, the SS7 link finds an open path through Chicago, Illinois. After it finds an open path, the SS7 signal secures the path by reserving timeslots for the voice path to travel. After SS7 establishes the path, your call takes the road that was built, and the call is completed quickly.

All the work that SS7 does prevents your call from having to worm its way to the destination through primary and secondary route paths, so your connect time is much faster. The SS7 protocol also shortens the time required to tear down the call after you hang up, so it's more efficient on the front and back end of the call. Did I mention that it's fast?

Ordering up SS7 service

If you are interested in ordering SS7, you should first call your hardware vendor to determine what it takes to integrate your system into an SS7 environment. You then need to find an SS7 provider. Your carrier can give you an SS7 connection to its network, and order the SS7 link to make it all happen, but you still have to be able to accept the SS7 configuration on your phone system. Unless you want to order more than $20,000 in hardware and keep an SS7 tech on staff at all times, you'll need to hire an outside company to care for your SS7 connection. It's best to begin the search for an SS7 provider by getting suggestions from your carrier.

After you find an SS7 provider, the SS7 provider needs to give you some information so that you can pass it on to your carrier and complete the order:

- **The name of your SS7 provider:** In order to coordinate your order for the SS7 link, your carrier needs to know who you are doing business with. This should be the easiest piece of information you have to provide.
- **The CLLI code of your SS7 provider:** The (pronounced) code identifies the physical address (in telecom terms) of the building that holds the SS7 provider's hardware. For example, a CLLI code of `LSANCA01` indicates that your SS7 provider's hardware is located in the 1 Wilshire Building in Los Angeles, California. The SS7 provider should have no problem providing the information.
- **The point code of your SS7 provider:** Your SS7 carrier's *point code* specifies the piece of hardware that will be receiving the SS7 signaling information. Again, this is information that is readily available to the SS7.
- **The TCIC range to use:** The TCICs (pronounced *tea-kicks;* stands for Trunk Circuit Identification Codes) range identifies the individual channels of your T-1 or DS-3 that will be using the SS7 link.

After you have this information, you can submit your order, wait for your SS7 provider to install and test the SS7 link with your carrier, and then cut the service onto your circuit. There are mountains of information printed about SS7 and the specifics of the signaling protocol, but as long as you know the previous information, you can order the circuit. After it's installed, just keep the SS7 provider's number handy in case there is a problem and you need to troubleshoot.

SS7 is a mature protocol used within the world of telecom and has evolved as a very stable and dependable way to transmit calls. If you would like more information about SS7 signaling, please check out `en.wikipedia.org/wiki/SS7` for an in-depth review of the protocol and additional links to even more information.

Feature Group D

Feature Group D (FGD) is an older protocol that is generally used only when a hardware limitation prevents a business from receiving ANI Infodigits on inbound toll-free calls. (Please scan the section in Chapter 5 regarding pay phone surcharges if you are scratching your head about the true identity of ANI Infodigits.) Your phone system might not be able to handle ISDN, or your carrier might have a limitation in its overhead that prevents it from sending ANI Infodigits. This is the only benefit of FGD, because it's an older protocol than ISDN, and it doesn't use the standard touch-tone DTMF (Dual-Tone Multi-Frequency) tones to send information. Instead, it uses an older multi-frequency (MF) signaling that takes longer to send and decode. A simple domestic outbound call is sent through the FGD protocol as:

```
KP + II + ANI + ST + KP + 7/10 Dig + ST
```

In this equation, the abbreviations represent:

- **KP:** Kick pulse that initiates an FGD transmission.
- **II:** Infodigits (identifies the origination type of the call, such as pay phone, hospital, prison, and so on).
- **ANI:** The origination's phone number.
- **ST:** Stop tone that acts as an end notification in FGD.
- **7/10 Dig:** The phone number you dialed.

The sequence and length of the individual sections of the transmission can vary, along with the sequencing of some portions of the information stream. The MF signaling requires more time, so a simple domestic call might take two or three times longer to send and connect using FGD than when using loopstart.

The overall age of the protocol and the many variations possible make FGD a challenge. The sequencing for inbound calls on toll-free numbers can also vary, so your hardware vendor must be familiar with your equipment before you begin using FGD. Problems can persist on your circuit for days and weeks if your hardware vendor is trying to solve the problem and master the protocol at the same time.

Feature Group D is a useful protocol because it allows you to receive ANI Infodigits, but it's more difficult to work on and it's slower, so I wouldn't recommend picking it as an option, and if you already have it, I would recommend moving on to something a little less cumbersome. I suggest you use ISDN if you need to capture the ANI Infodigits on your inbound toll-free calls. The hardware is easier to come by, as are technicians well versed in the protocol.

Understanding Local Loop Pricing

After you have all the configuration information to order your circuit, you should turn your attention to your local loop. The term *local loop* refers to the copper wire or fiberoptic cable connecting your building to the switching hardware of your carrier. The reason it's called the local loop is because a local carrier almost always delivers it. Even if you order a dedicated circuit from your long-distance carrier, the carrier will still subcontract with a local carrier to deliver the circuit to your hardware because it saves lots of money by doing so.

Even the biggest carriers don't have a battalion of technicians in every town in the U.S. They also don't have an existing network of cabling that connects every business and home in every city. The good news is that the local carriers are perfectly set up with the network and technicians to install local loops. Because the long-distance carriers don't want to rebuild the wheel when it comes to local service, they simply contract with the people who are already established.

A *POP*, or *Point of Presence*, is a building, or a suite in a building, where your long-distance carrier houses its network hardware. If you wanted to physically lay your hands on the massive computer system that routes your calls and generates call logs that spawn your billing, you need to visit a POP. A *CO*, or *central office*, is the local carrier's POP equivalent. The local loop of your dedicated circuit terminates at the POP of your long-distance carrier (if you're ordering a long-distance circuit) or at the CO of your local carrier (if you're ordering a local circuit).

The first point to determine about the local loop is its cost. Your carrier should provide you with a price quote for the loop, typically when you negotiate the contract. If the contract procedure takes more than 30 days, be sure to get a new price quote before you place the order. The cost for local loops fluctuates over time and you might be looking at a loop that is $10 to $20 less than the price you negotiated, or $100 more. It's best to validate the price before you get locked into a contract for 12 months. If the local loop is too expensive, there are cheaper alternatives.

Understanding the elements of local loop pricing

The cost of a local loop is based on three primary elements:

- **The cost to terminate the circuit into your carrier's network:** This is frequently called a *chan term*, short for *chan*nel *term*ination. Basically, you pay for the connection between your long-distance carrier and the local carrier. If you order a circuit to your local carrier, you still pay for a chan term, even though your local carrier is just terminating the circuit to itself. Blame it on the profit assurance department working overtime.
- **The mileage between the POP or CO and your building:** Carriers use this equation to determine this charge:

  ```
  Rate x Mileage per channel x 24 = Local loop mileage
         per T-1
  ```

  Basically, the rate is determined by the mileage and based on an individual phone line. Because most people buy T-1s, the mileage then has to be multiplied by 24 because a T-1 is equivalent to 24 individual phone lines. The mileage is calculated based on *airline miles* (your sales rep may say that the distance is measured *as the crow flies*) between your building and the carrier's POP, not based on the distance it takes to drive between the two locations. That's nice if there's a big, impassable mountain between your corporate headquarters and the POP.

- **The cost to terminate the circuit into your building:** This is the second chan term that you are charged for. Whenever a dedicated circuit connects into a carrier, either the company delivering your circuit into your carrier, or your carrier who contracted with someone to pull the cable from your office, is going to charge a monthly fee for the connection.

Investigating your pricing options

If you receive local loop pricing for a dedicated circuit that seems a bit high, you have other options. You need to keep the local loop fee in mind when you make your decision about which carrier to use. If one carrier charges an average of 2.5 cents per minute for your calls, but its local loop fee is $1,250 per month, it might be cheaper to use another carrier that charges you 3.5 cents per minute but charges $250 monthly for its local loop. You simply have to crunch the numbers to determine which scenario is cheaper. If you make 100,000 minutes worth of calls, you will save the extra $1,000 on the difference between the two rates. If the scenario is different and you only have 40,000 minutes, the $1,250 local loop fee ends up costing you $600 more at the end of the month.

When it comes to local loops, you do have a few options to reduce the cost:

- **Find a closer *Point of Presence (POP)*:** Not only does your physical location affect the cost of the local loop, but so does the physical location of the connection to your carrier. If your business is located in a big city, there might be multiple POPs into which your local loop can terminate. Usually, the automated pricing system for your carrier chooses the closest POP, but it isn't 100 percent accurate. If you have a ridiculous quote from your carrier, you can ask the carrier to check out other POPs in the area.
- **Order the local loop yourself:** You can call your local carrier, or other competing local carriers, to receive quotes for the local loop. Your long-distance carrier negotiated local loop pricing with the local carriers quite a while ago. If your local carrier is running a promotion or special deal on local loops, you might get a better deal if you order the local loop yourself. You have to coordinate the connection between your local carrier and your long-distance carrier's network, but this little bit of extra work can save you money in the long run.
- **Move your office closer to your carrier's POP:** This might sound like an extreme measure, but if the main expense for your business is your long-distance bill, it might be worth it. Depending on the nature of your business, you could even go so far as to rent a suite in the same building as the local carrier's POP.

Evaluating Your Local Loop Choices

You have three basic options for how you order the local loop portion of a dedicated circuit. Each option has different benefits and drawbacks. The best choice for your company depends on how much your business is based on phone service, and the level of responsibility you want for ordering and troubleshooting your circuit.

The three basic options for your local loop are:

- **Have your carrier order the local loop.** If you don't have an in-house technician and aren't very comfortable with telecom, this is your best choice. Having your carrier order the loop reduces your responsibility for coordinating the order and troubleshooting the circuit.
- **Order the local loop yourself.** If you are technically sharp and want to be responsible for all aspects of the local loop, this is your best choice.
- **Eliminate the local loop by collocating to the same building as your carrier.** Companies typically take this option when telecom is their business, and their employees work to take care of their phone system, rather than vice versa.

Your interaction in the order process and the average time to completion will vary depending on the option you choose. The same steps happen during every ordering process, even though some of them may happen without you knowing about it.

Understanding the complexities of type 2 (and type 3) circuits

In a perfect world, you order a dedicated circuit from your long-distance carrier, the long-distance carrier contracts with the local carrier in your area, and the circuit is delivered. Local loops set up like this, with only one local carrier, are called *type 1 circuits*. Unfortunately, this isn't always how things play out. Sometimes, the territory for your local carrier stops short of the POP where your long-distance carrier is located. Your local carrier then must contract with the neighboring local carrier to connect to the POP in what's called a *type 2 circuit*. It can get even worse if your long-distance carrier doesn't have the facilities available to receive the local loop from the second local carrier, and has to contract with a third local carrier (*type 3 circuit*).

All these circuit types are relevant to you because every time you add another local carrier to the loop, you can expect to add 15 days to the installation date of your circuit. Every carrier adds more possibility for error in provisioning the circuit, because the order now must run through another company's order system. And, of course, there's always the ka-ching factor: a higher local loop cost per month, and more difficulty if you ever have to troubleshoot the circuit. If you can avoid a type 2 or 3 circuit, it's in your best interest to do so.

Speeding Up Order Processing?

If your order is moving slowly, or if you're under a time constraint, you might be tempted to ask the carrier to expedite your order. Although this idea seems good in theory, you need to ask two questions before you commit to expediting an order:

- **How much is the expedite fee?** The cost to speed things up can be anywhere from $500 to $5,000 per order, or per circuit. After you know what the expedite fee is and the daily savings you expect when it's installed, you can ask the second question.
- **What exactly do you get when the carrier expedites the order?** Many carriers won't guarantee you anything more than increased visibility on an expedited order. The manager or director of the provisioning department is now aware of the order's existence, and may move it to the top of the pile for provisioning and design, but that's hardly worth $5,000. If you have no guarantee that the circuit will be installed any quicker than it would be if you simply pick up the phone and make a nagging phone call to the carrier every two days, don't spend the money. A little diplomacy and persistence on your part can give the order the same visibility by the upper management, without paying the expedite fee.

If you are ordering a long-distance circuit, the fee for expediting an order doesn't generally cover expediting the process at the local carrier level. If you want a circuit expedited at that level, you have to open up your checkbook — wide. Fees for expediting the process with local carriers are generally much higher than those charged by long-distance carriers; some carriers even ask for a blank check so they can tell you the fee after the work is done. Like long-distance carriers, you're given a rather flimsy guarantee that the order will be pushed along, but you won't see a commitment that your circuit will be installed a week, a day, or an hour earlier than the carrier's original estimate.

The one point that both local and long-distance carriers commit to when you ask for an expedited order is the fact that you will be charged for it. Will the order actually be done faster? Who knows? All bets are off.

Ordering a Carrier-Provided Loop Circuit

The simplest order you can place for a dedicated circuit is one where your carrier orders the local loop. After you fill out all the technical paperwork and submit it, on average, the circuit is available for use about 30 to 45 days later. Allow the order to chug through the system for three days before you call your carrier to get an order number for the circuits. If, after three days, the local carrier hasn't generated an order number, the order might have gone astray. You should correct this problem now instead of starting the process over on day 25.

Getting realistic when your order becomes a project

Set realistic expectations, and the second you hear someone say the word *project,* take a deep breath. All local and long-distance carriers have categories for their orders. As an order becomes more complex — see the sidebar, earlier in this chapter, "Understanding the complexities of type 2 (and type 3) circuits" — it can be classified as a project. When an order becomes a project, this is a Bad Thing, because the new title places the order under greater scrutiny. More scrutiny means more layers, and more layers mean more processes. More processes mean more time. If your order ends up being handled as project, add 15 days to your expected due date.

Receiving a Firm Order Commitment

After you receive your order number, you should check on the order about once a week to make sure that it's on track. At this point, you are waiting for the *Firm Order Commitment (FOC)* document; it tells you when your local loop will be installed at your office and tested between your carrier and the local loop provider. There is no industry standard for the design of the FOC (pronounced either *eff-oh-see* or *fawk*) document, but it should list most of the following information:

- **The long-distance carrier circuit ID:** This is the circuit ID written on the circuit after the cables for it are installed at your building. This number is the one piece of information that will be logged in at your carrier's customer care center. If your circuit has a problem, you need this number so that you can identify your circuit and troubleshoot it. Keep this information readily available.
- **The order number used by your long-distance carrier:** After the wiring installed at your office is connected and you activate the circuit, you should mention this order number to cross-reference the circuit ID and make sure everyone is working on the right T-1s.
- **The trunk group name(s) for your dedicated circuit:** You need to know this information if you ever add toll-free numbers to your circuit. Many toll-free features are commonly installed in the trunk group, so your carrier won't care about the circuit ID when you are adding toll-free numbers. This info is also good backup to help your carrier find your circuit if no one can find the circuit ID.

- **The billing ID code for your dedicated circuit (this may be your trunk group name, or referred to as an Auth Code, CLLI CDR, or DL value):** Keep a list of these codes; if one appears on your bill that you don't recognize, call the carrier to check on it.
- **The name of the local carrier:** Because your local carrier was contracted by your long-distance carrier to help out, you should determine who did the work in case you have a problem with your circuit. If your long-distance carrier says it has isolated the issue to your local carrier and has opened a trouble ticket with Bell South, but your FOC document says the local carrier is Time-Warner, you can correct this issue quickly. It's better to validate this information five minutes into a trouble ticket rather than after you have lost five hours.
- **The local carrier circuit ID:** Match up every local carrier circuit ID with every circuit ID from your long-distance carrier. If the local carrier has to be dispatched to work on your T-1s, the technician can instantly know whether he or she is looking at the right one.
- **The plant test date (PTD):** The *plant test date,* or *PTD,* represents the date on which the local carrier plans to test the continuity of the circuit with your long-distance carrier. This is generally about 24 hours before the committed due date.
- **The committed due date:** This is the date on which the local carrier agrees to have the circuit released to you for testing.

The date you receive the FOC document can be anywhere from two to ten days before the wiring for the local loop is actually installed at your office and tested for continuity. The term FOC is used so casually in the world of telecom that people have blurred its meaning. Don't expect to get your circuit on the date you expect the FOC document to be delivered. It can happen, but it isn't the industry standard.

Ordering a Customer-Provided Loop Circuit

A *customer-provided loop* identifies a dedicated circuit where you order the local loop instead of having the carrier do it. You might have found a better rate for the local loop through your sales rep, or you may simply want more control over the order and the troubleshooting process. As you fill out the order form for your carrier, you need to note the section devoted to the local loop because it has about four questions that you need to answer. Your carrier needs to know the following:

- **The name of the local carrier you are using**: The local carrier you choose might not have an existing relationship with your long-distance carrier at the POP specified. Every long-distance carrier has only a limited number of cables allocated for connections with each local carrier. Your long-distance carrier could be out of available cable pairs for your proposed local carrier. You need to wait for new cables to be installed or choose another local carrier to connect into your long-distance provider.

Your long-distance carrier might not be able to use the local carrier you have chosen. If you aren't using a Baby Bell company or a standard local carrier, your long-distance carrier might not have any contracts, arrangements, or facilities allocated for the company you want to use. Before you set your heart on using a local carrier to make the connection into the long-distance network, confirm that your long-distance carrier can use this company.

- **Contact person at the local carrier:** Someone at your long-distance provider may want to chat with someone at the local carrier you chose to work out the finer details of the circuit. Long-distance networks generally make you do all the coordination, because, after all, this is a customer-provided loop.
- **Contact phone number for your local carrier:** If your long-distance carrier is going to contact your local carrier for any reason, it needs a phone number. Don't expect the carrier to use it and cut you out of the process (see "Ordering a local loop makes you responsible for it," later in this chapter). The long-distance carrier probably wants the number on file if there is ever any trouble on the circuit.
- **POP address or CLLI code:** This is very important information. Carriers have POPs throughout the U.S., and some urban areas have multiple POPs. If you're looking for a Qwest POP in Atlanta, Georgia, for example, there are at least two locations you can use.

When you list the physical location where your office is located, your carrier provisions the loop out of the POP closest to that premises. If two POPs are of equal distance from your office, you have a 50-50 chance that each carrier will choose a different location. The next thing you know, your long-distance carrier has built a nice new circuit from the Atlanta 2 switch and patched it into its network; meanwhile, your local carrier has installed a nifty circuit to the Atlanta 1 switch. The disjuncture can eventually be corrected, but you will have lost several weeks of provisioning time.

Request a *POP list* from your long-distance carrier before you order, or ask for a price quote for your local loop. The document lists every POP your carrier has, along with its physical location and the industry CLLI code. The *CLLI code* is telecom shorthand for the physical location of the POP — every carrier can identify the POP by this code. Along with the address and CLLI code, the list should also have a phone number for each POP that your local carrier can use to determine the cost of the loop.

Ordering a local loop makes you responsible for it

Only the company that is the *end user of record* is authorized to initiate trouble reports on a dedicated circuit. If you order a local loop, your long-distance carrier can't (legally) initiate trouble reports with your local carrier for your circuit, or speak to your local carrier directly about the specifics of your circuit. For this reason, you should order the local loop only if it makes a large financial difference, or you are technically savvy.

At the end of a frustrating day, a carrier may abuse the fact that it isn't the end user of record to pawn off a problem to you. This can devolve into childish finger-pointing as each carrier maintains that a problem is someone else's responsibility to fix. For this reason, it's easier to pay a little more money and have your long-distance carrier order the local loop. Then if there is ever a problem on the circuit, it doesn't matter where the problem is located; your long-distance carrier still has to fix it because it is the end user of record.

If you're still reading this section, then you haven't taken my advice and are ordering your own local loop. That's fine, too. After you have filled out and submitted the order forms for your long-distance and your local carriers, you have to let them do their work. Give them both about three days before you follow up and ask for order numbers. As long as they each have order numbers for your job, everything should be flowing fine.

Understanding the CFA

About 15 to 25 days after you place an order for a local loop, you should receive the *Carrier Facilities Assignment,* (CFA) document from your long-distance carrier. This document identifies the exact piece of hardware to which your local carrier is to deliver your circuit.

The document your long-distance carrier sends you should actually be an *LOA CFA (Letter of Authorization Carrier Facilities Assignment).* This document doesn't just tell you the specific piece of hardware to which your local carrier is going to connect, but also authorizes your local carrier to enter the POP to leave cables for your long-distance carrier to complete the final connection into their hardware. The CFA lists the following information:

- **The local carrier to which it has assigned the CFA:** Look at this section closely. If your long-distance carrier has the CFA directed to ICG Communications and you are using Verizon, you might need to go back and ask for a new CFA.

- **The physical location of the POP, or the POP CLLI:** This address or CLLI code should match the address you sent to your local carrier when you ordered the loop. If the address isn't an exact match, have both carriers clarify the location.
- **The circuit ID of the circuit:** The circuit ID on your CFA is the number that your long-distance carrier will use to reference this circuit for as long as you have it. You need this number if you make any changes to the circuit, or if you have to open a trouble ticket on it. Keep this information handy at all times.
- **The bay, panel, and jack into which your circuit will connect:** This line item of the CFA is the most important. As long as you have this line item of text, you can send the order to the next step with the local carrier. The specific information may look like gibberish to you, but to a technician with your local carrier it functions like a set of GPS coordinates.
- **A small disclaimer that the CFA will expire in 30 days if it's not used:** Nothing is forever, and the saying holds true in telecom as well. If you don't use your CFA in 30 days, it might expire and be offered to someone else. This is important to remember when your local carrier is pulling cable or installing new hardware before they can make the connection, causing delays in your order. If you're approaching the 30-day mark, try asking your carrier for a few more days. If you're looking at a problem that is going to take 60 days or more to correct, you have to request a new CFA when it's closer to your expected resolution.

The CFA is sent to you only if you ordered the local loop. After you receive the CFA, you should check it out, make a copy, and forward the document on to your local carrier immediately unless there's a problem. The local carrier uses the information in the CFA to design its portion of the circuit. Your local carrier can't do any design work on the circuit until it receives the CFA, so it's important to get the CFA to the local carrier as quickly as possible.

Understanding the Design Layout Record (DLR)

Your local carrier will take 2 to 15 days to design and construct its portion of the circuit from the CFA you supplied. The time frame varies quite a bit with all carriers. Smaller carriers can finish a design within 48 hours, but larger companies have set procedures that lengthen the time it takes to do the same thing. After the circuit is designed, your local carrier will send you a Design Layout Record (DLR) document that outlines the circuit and all the pertinent technical information about the local loop.

At first glance, the entire DLR document might look like nonsense. If you look hard enough, though, you find at least the following information scattered somewhere on the DLR:

- **The order number used internally by your local carrier:** Your local carrier generates its own internal order number for your local loop for tracking purposes. The local carrier might also list the long-distance carrier's order number, but that is simply as a cross-reference.
- **The circuit ID used internally by your local carrier:** This is the code by which your local carrier refers to your circuit from the day you receive your DLR until the day the circuit is cancelled. You need to keep this number handy in case you change the circuit or have trouble with it. The local carrier's customer care rep will ask you for the circuit ID in order to open up a trouble ticket.
- **The CLLI code for the first local carrier CO to which your circuit terminates:** This code identifies the physical location of the local carrier Central Office of your circuit before it reaches the POP of your long-distance carrier. This enables your long-distance carrier to identify with greater accuracy the cables that are placed at the CFA location.
- **The CLLI code for your long-distance carrier's POP:** Confirm that the CLLI is correct and that your local carrier hasn't decided to take your local loop somewhere it was not invited.
- **The bay, panel, and jack from which the cabling will leave your local carrier's network:** This is important information, because your long-distance carrier may need to validate the cabling provided by your local carrier to ensure it is plugging into the right CFA point.
- **The plant test date (PTD):** This is the date your local carrier is planning to test the circuit with your long-distance carrier.
- **The committed due date:** This is the date your local carrier expects to have the circuit complete and ready for you to activate.

After you receive the DLR from your local carrier, you need to forward it to your long-distance carrier. The long-distance carrier uses the information to identify when your local carrier is going to finish its end of the circuit and gives the long-distance carrier some indication of how the order numbers, circuit IDs, or customer name may be written on the cabling provided by the local carrier.

Using a meet-me room (MMR)

All large carrier hotels have a meet-me room that is either a suite or a complete floor in the building. Abbreviated as MMR, this room acts as a general cross-connect area for all carriers. The meet-me room looks like a zoo for computer equipment with rows of locked cages containing servers and switches of every carrier (local and long distance) as well as every collocation provider in the building, all flashing green, yellow, and red lights. If your carrier is on the 13th floor and your collocation provider is on the 2nd floor, you may be able to avoid paying for a zero-mile loop if both companies also have a cage in the meet-me room. In this case, you should request a CFA from your carrier at the meet-me room and let your collocation provider finish the cross-connect. These are frequently the easiest and quickest connections you can make to install a circuit.

The final connection of your circuit between your carriers is completed in steps:

1. The technician from your local carrier physically drops 6 to 9 feet of cable to the CFA point.
2. The technician locates the cage identified in the CFA document, writes down the circuit ID or customer name on the cable to tag it, and then feeds down enough cable for the long-distance carrier to finish the connection.
3. The technician from your long-distance carrier looks for the cable in its cage based upon the information on the DLR and connects the circuit into the assigned CFA point.

Ordering a Circuit Without a Local Loop

If your business provides calling card services or telemarkets customers by leaving a prerecorded sales pitch that sounds like a personal message on people's answering machines, you're a great candidate for a direct connection into your long-distance network. The greatest portion of the monthly recurring charge on a local loop is the mileage fee. To reduce the cost, you can try to find a POP closer to your office, or you can move your office closer to the POP. The ultimate cost saver would be to move your office into the same building as your carrier so you don't have to pay for any mileage.

You don't need to move all of your staff into your carrier's POP — just your phone system. As long as you can access your hardware remotely through either an Internet connection or a simple modem dialup, and your business model can support this scenario, you can save money every month in local loop fees. Maybe once every few months you'll need to visit your phone system to give it a software upgrade or some maintenance, but as long as you don't need to have access to it every day, this is a great option. Fortunately for you, there is an entire industry created to help you.

Introducing the carrier hotel

Every large, metropolitan area has a *carrier hotel* that houses POPs and COs for almost all carriers, both long-distance and local. These buildings are meeting centers for telecom connections; hundreds of thousands of inter-carrier connections are made in them. Larger cities have several smaller carrier hotels in close proximity to the established primary building. The carrier hotels not only house carriers, they also are the home of *collocation companies* (also called *colo providers*), which exist only to give cheap carrier access to companies like yours.

It isn't a good business decision to rent out an entire suite in a carrier hotel. You don't need all the space and you will still have to run cables to the floor of your carrier, build a temperature-regulated server room, and install security measures so people can't access your hardware. In the end, all you need is about 9 feet of rack space and some electricity, not 800 square feet and a key to the washroom. The colo provider's job is to provide electrical power and enough floor space in their suite to install your hardware. The colo provider probably provides you with either a secure cage or cabinet for your hardware. This is the bare minimum that colo providers offer. Some may provide maintenance contracts for your hardware, cabling service to connect you to your carrier, and varying levels of climate control and battery backup for whatever you install.

If you are planning on moving your phone system to a carrier hotel, think nationally and factor in your rates and the taxes. A colo provider in Alabama may have a better rate and the same access to your carrier that a carrier hotel in Las Vegas offers. If most of your calls don't terminate in Alabama, you could save a lot of money as you are now charged an interstate rate that is generally cheaper. You may also look at the tax rates assessed in different states. The state tax on a colo space in Las Vegas, Nevada, is considerably less than on a colo space in Los Angeles, California.

Selecting a colo provider

After you settle on the carrier hotel to which you want to move your telecom hardware, you need to find a colo provider in the carrier hotel that will house the hardware. If you haven't been working in telecom at that specific carrier hotel, you probably don't have any idea which colo provider is the best. There are two resources that can help you with this quest: your hardware vendor and your long-distance carrier.

Your hardware vendor may have worked in the building before and has contact with specific colo providers. If the rep from your vendor doesn't know of a colo provider there, perhaps you can ask the rep for referrals from other vendors. If you are using a carrier hotel outside your state, your hardware vendor might not have any leads for you. Before you turn to the Internet and begin searching for colo providers, you should call your long-distance carrier, who can also provide some referrals from its technicians who work at the carrier hotel. A preexisting relationship between your carrier and the colo provider makes the move that much smoother.

Every colo provider doesn't have direct access to every long-distance and local carrier. Some may specialize in connections to MCI and Broadwing only, or a handful of local carriers. If you were referred to a company by anyone other than your long-distance carrier, confirm that the colo provider can connect you to the right network.

Colo providers have a variety of charges they can assess to you, aside from rack space and power. You will see fees for any maintenance or troubleshooting that you want the provider to perform. Generally, the fee increases for after-hours service, so the midnight installation you have planned on Saturday (to keep your business running seamlessly) will cost twice as much as it would if you did it at noon during a business day. Of course, if you lose three times as much business because your system is down in the middle of a busy workday, you may be more than happy to pay the increased fee for after-hours work. Check the complete list of charges for all of your prospective colo providers before you decide on one.

Ordering the cross-connect

A *cross-connect* is the wiring that joins one carrier to another. If you are located with a colo provider in the same carrier hotel that your long-distance carrier provided CFA for, you don't have a traditional local loop spanning between your multiplexer and the switch of your long-distance carrier. All you have to worry about is having the cross-connect installed. The process for ordering a circuit without a local loop is almost identical to ordering a circuit with a customer-provided loop (see "Ordering a Customer-Provided Loop Circuit," earlier in this chapter). The main difference is that instead of passing CFA information to a local carrier, you send it to your colo provider.

Your long-distance carrier still needs to give you a CFA that references a section of their hardware allocated for your local carrier. If your colo provider provides the cross-connect, you may not be allowed access to the main suite of your long-distance carrier, and the connection has to be made in a meet-me room in the building. If the CFA provided was on the tenth floor of your long-distance carrier's POP, and now must move to the fourth floor meet-me room, you may see some challenges. The cost of the cross-connect fee may be more for a meet-me room than it is in the long-distance carrier's suite. You may experience a delay while the circuit is being designed with a new CFA on the fourth floor. This effort may all be for nothing in the end because your long-distance carrier may not have the CFA at the circuit level you require. You may need a pair of T-1s, but your carrier can only provide DS-3 or OC-3 connections.

Because colo providers are smaller than local carriers, they can usually act on your CFA and return a design layout record (DLR) in about 24 to 48 hours. That's a pretty short time frame when you consider that a larger local carrier can take weeks to dispatch a technician to your building to install the circuit. If you have a long-standing relationship with the people who establish orders at your long-distance carrier, you might be able to install a dedicated circuit to a colo provider in as little as two weeks.

Understanding the zero-mile loop

The entry points to a carrier's network are very specific. Instead of going by the street address of the building, the entry point is narrowed down by the floor and suite within the building. For example, if your colo provider (on the 2nd floor) doesn't have access to your carrier (on the 13th floor), someone is going to pull cable 11 floors and into the suites. Your carrier can run the wiring for you. All you have to do is ask for CFA on the second floor in your colo provider's suite. Because your carrier is coming to the colo provider, and not the other way around, the carrier charges you an installation fee and a monthly maintenance fee, just as if it were pulling cable to a building across the street or 10 miles away. When cable is installed within such a small space, it's called a *zero-mile loop*. A zero-mile loop is a local loop that doesn't leave the building.

If you set up a zero-mile loop, I recommend that you ask for a CFA from your carrier and pay your colo provider to run the cabling; this option is generally cheaper and quicker.

Preparing for the Installation

Before your circuit is released to you, your long-distance carrier has to test it to confirm that none of the wiring or cross-connects were forgotten when the circuit was being installed. The final test ensures electrical continuity on the circuit from the long-distance carrier's hardware to the end of the local loop where your CSU or multiplexer is plugged into it. After the carrier signs off, you need to schedule installation of the circuits. However, before you do that, there are two other issues you need to address: the inside wiring and your cut sheet.

Inside wiring is just as important on a dedicated circuit as it is with a single fax line. You need to ensure that all the required cables and wiring are pulled to the room that holds your phone system. If you have any doubt about what needs to be done to complete the inside wiring of your circuit, check out the inside wiring section of Chapter 4.

Creating a Technical Cut Sheet

Information is rarely where you want it when you need it the most during an emergency. This is why you should draw up a technical cut sheet for every dedicated circuit before it is installed. One week after the circuit is activated, your CFAs, DLRs, and FOCs might end up in the basement of your building. If the circuit fails, you have to run down and dig through the files to find the information.

The Cheat Sheet at the beginning of this book contains a starter cut sheet to help you organize the volumes of information about your dedicated service, installation, features, hardware, and contacts.

Wouldn't you rather have all this information gathered in one document so you don't have any confusion over who to call or what your circuit IDs are? The cut sheet is typically a simple Excel spreadsheet with the following information listed:

- **Physical location of the circuit:** You might have hardware in multiple locations that you are responsible for maintaining. Without having the address, you can end up spending two days troubleshooting the wrong circuit.
- **Carrier:** You may have multiple carriers, as well as local and long-distance circuits. An accurate report of which carriers you have for each circuit prevents confusion when you issue change orders or open trouble tickets.
- **Trouble reporting number for your carrier:** Every carrier has its own toll-free number for trouble reporting. If you don't have to call the number every few days, you will forget it when you need it the most. If you keep the number on your cut sheet along with the other information, you won't have to chase it down.
- **Circuit ID(s):** These are the most important pieces of information you are given. Any activity that needs to be done on your circuits will be referenced to either the circuit ID or trunk group name.
- **Trunk group name:** If your carrier is having difficulty finding the circuit ID in its system, the trunk group name is the next best piece of information available to clarify things. Not every document is 100 percent accurate, and if your FOC or CFA was transposed when it was given to you, the trunk group name can help your carrier find your circuit when you are adding service or troubleshooting.
- **Account number:** Some carriers require you to submit your account number on any document you send that requests a change to your service. Some carriers issue unique account numbers depending on the service you order from them, so you may have multiple account numbers for a single carrier. Whatever the case may be, make sure that you list all account numbers (associated with specific services, of course) in one place.

- **Order number:** If you are trying to place an order for a change of service a circuit and your carrier can't find the order by either the circuit ID or the trunk group number, try using the order number of the initial installation — it's a pretty solid backup for tracking down the circuit. I would like to say that you'll never have to go this far to help a carrier find your circuit, but it does happen at times.
- **Local carrier circuit ID:** You should be able to link every circuit ID from the local carrier that installed your loop to the associated circuit ID for your long-distance carrier. This is an especially important thing to do when you have an issue with your long-distance dedicated circuit and a technician from the local carrier is dispatched to correct it. The circuits at your office may only be tagged with the local carrier circuit IDs. If this is true, you need to know which local-circuit ID matches up with which long-distance circuit ID or you might spend time working on the wrong circuit.
- **Local carrier trouble reporting numbers (611, toll free, and local number) if you ordered the local loop:** If you are the one who ordered the local loop portion of your dedicated circuit, you *need* this number. Ask for a toll-free number and a direct number for the repair department of your local carrier. The division of your local carrier that handles the troubleshooting of dedicated circuits isn't a standard option in the carrier's general directory. You may need to speak to someone in a special division of business service, so you should get the number before you install your circuit.

  If your long-distance carrier ordered the local portion of the circuit, the local carrier won't speak to you about troubleshooting issues.
- **Configuration (line coding/framing/outpulse signal and start):** When you test a dedicated circuit, you should, of course, identify it by its circuit ID and trunk group, but also by its configuration. If you have two circuits that use ISDN, and two that are plain-vanilla in-band E&M Wink, you want to mention which protocol is associated with which dedicated circuit.
- **Trunk group configuration:** Identify the hunting sequence of your incoming calls, and how many T-1s or individual channels are in each trunk group. The configuration of your circuits may change over time as you upgrade your hardware, add more toll-free numbers, and your company evolves. You should know the most recent incarnation of your circuit so you don't submit change orders based on a configuration that was changed six months ago.

  Of course, this means that you need to update your cut sheet every time service, configurations, and protocols change.
- **Provisioning contact names and phone numbers:** If you need to correct a provisioning error during the installation process or have to submit a change order, these numbers are essential. The provisioning staff is also

useful if you have a trouble issue that you need to escalate to a higher management level.

- **Any special features on a trunk group:** If you have a block on your dedicated circuit to prevent outgoing international calls, or if your carrier has applied special routing features on a specific trunk group, you need to identify the trunk group with the special features. Later, if you need to place an order to remove the international block or other feature, you won't want to spend five days tracking down which trunk group has to be opened.
- **The date the local loop was installed and completed by the local carrier:** This date tracks the life of your circuit so you know when the contract term expires. The gestation period of a dedicated circuit may be 45 to 60 days before it is ready to process calls. This time is factored into most contracts for dedicated service so that the 12- or 24-month term actually begins on the date the local loop is installed and is ready to activate, not on the date the contract is signed. Circuits could be delayed by six months before they are installed, and your long-distance carrier doesn't want to be halfway through a contract period before seeing any traffic on the circuit.
- **The date the circuit was installed and accepted:** This date tracks the circuit and helps identify the date on which your calls should start appearing on bills from your new carrier. If you have outbound calls on your new dedicated circuit prior to that, you need to investigate further.
- **A list of any toll-free numbers on the circuit:** Be sure to keep a complete list of all toll-free numbers, and make sure the list links the toll-free numbers with the trunk group (or circuit ID) to which they terminate. Also have information about any overflow configurations that you've set up for them.
- **A list of special features on toll-free numbers, such as DNIS, ANI Infodigits, and ANI delivery:** If you order another circuit to add on to your existing trunk group, you need to know the configuration so that you can ensure the toll-free numbers coming in on the new span receive the same treatment.
- **The fees for your local loop or cross-connect:** These fess include both *monthly recurring charges* (MRCs) and installation fees. When you reconcile the invoice from your carrier for your circuit, you need to validate everything you're being charged. It is much easier to enter a list of fees into one central document than to try and track them down 45 days later when you receive an invoice.
- **The hardware vendor's name and contact information:** Any changes you make to your circuit may require your hardware to be reprogrammed. You may also need to contact your hardware vendor if your carrier has isolated an issue to your phone system. There are some phone numbers that you can't have listed on too many pieces of paper, and this is one of them.

- **A section for notes:** Every circuit has its idiosyncrasies. Be sure to make note of anything that you believe might helpful at a later date. This section might include the name and phone number of the technician at the carrier that installed the circuit, or any confirmation numbers associated with the installation. If anything was done out of procedure, note it here. For example, maybe a protocol setting or other configuration change (which would normally require a change order and ten days to process) was made on the fly because your carrier was nice enough to make the fix at the time of the installation. If something like this happens, by all means, write it down in your notes. Sometimes special favors aren't completely documented by the long-distance carrier, and the "corrected" (read "wrong") configuration may be reset to the requirements listed on the "initial" (read "wrong") order. Making note of who helped you and what was done will reduce circuit downtime later.

After you have transferred all the required information from the CFA and DLR documents (if you ordered your own local loop) or the FOC document (if your long-distance carrier ordered the local loop), your cut sheet is complete, and you are ready to schedule the installation of your circuit. If any of the information is missing from these documents, contact the appropriate carrier to fill out the information. Any information missing in their documentation may indicate that a portion of the circuit that is not complete, so validate everything.

Every carrier has its own procedure for installing circuits, from allowing you to dial up on the fly, to requiring you to schedule 48 hours or more in advance to secure a timeslot. The entire installation process is covered in detail in Chapter 10.

Chapter 9

Ordering Toll-Free Service

In This Chapter

- Reserving new and vanity numbers
- Moving a number to a new carrier
- Understanding the RespOrg document
- Handling toll-free rejections
- NASCing toll-free numbers
- Ordering and activating switched toll-free numbers
- Ordering dedicated toll-free numbers
- ANI II ordering and DNIS and ANI delivery
- Activating dedicated toll-free numbers
- Resolving activation issues

Toll-free numbers are so familiar that we take them for granted. The truth is that even though you may use them every day, you may not know how they work and how they flow through the world of telecom. That's okay. You don't need to know everything about how they work, but you do need to know a few things if you want to install toll-free service for your business.

A *dedicated toll-free number* points into a dedicated circuit at your office, whereas a *switched toll-free number* rings into a normal phone line like you have at your house. You go through the same steps to reserve or migrate toll-free numbers from one carrier to another.

The two toll-free phone options differ when it comes to the ordering process and the available features. This chapter covers everything you need to know about ordering dedicated and switched toll-free numbers; it also covers migrating toll-free numbers from one carrier to another, and addresses the interesting nuances of toll-free features. If you have general questions about what a specific toll-free feature is, or if it's something you might be interested in using, please check out Chapter 5.

Reserving New Toll-Free Numbers

Requesting a new toll-free number from your carrier is typically a quick and painless procedure. It might involve some very basic paperwork, or an e-mail might suffice. Some carriers have a reservation desk you can call into to reserve numbers. Regardless of the method for requesting new toll-free numbers, the carrier will ask you for a couple of details. Specifically, carriers want to know:

- How many toll-free numbers you want, and whether multiple numbers should be in sequence (1-800-SMILEY1, 1-800-SMILEY2, and so on).
- Whether you have special requests. If you need a special prefix (800, 888, 877, or 866?) or a vanity number (1-866-SMILEYS?), speak up now.

Requesting random toll-free numbers

Most carriers give you the option of requesting toll-free numbers by the first three digits, so if you do have a preference, you can specify whether you want an 800, 888, 877, or 866 number. After you place your request, your carrier dips into the national 800 SMS (Service Management System) database and reserve as many toll-free numbers as you need. (I talk about the SMS database in Chapter 5.) It's possible that some of your toll-free numbers will be in sequence, but it isn't very common. If you need a block of 10 or 15 *consecutive* toll-free numbers, you are more likely to find them in the 877 or 866 prefixes.

After your reservation is complete, the carrier sends you a list of the toll-free numbers assigned. You then have 30 days to activate those toll-free numbers before they are released back into the *spare* pool in the SMS database.

Requesting a vanity number

A *vanity number* is a lot like a vanity license plate: You know it when you see it. Any toll-free number that prevents you from simply requesting a random toll-free number is a vanity number. Often, vanity numbers have repeating numbers, a special number sequence, or spell out a product or company name.

Finding the right vanity number isn't always easy. Understandably, the best vanity numbers are already taken, so you may request a number only to find out that it's not available. In fact, all the good numbers in the 800, 888, and probably 877 are taken. The likelihood that you'll find an open number such as TAX-1040, FLO-WERS, REFI-NOW, or any other buzzword is very small. Because carriers have to check a national database before awarding you a number (and this is a time-consuming process), some carriers limit the quantity of vanity numbers you can request to search per day.

Dealing delicately with RespOrg departments

The people who work in a carrier's RespOrg department are generally well insulated from the rest of the world, and you may never speak to anyone who actually works in this department. I believe that many RespOrg departments were born out of reaction and quickly thrown together. Imagine the scenario: The carrier's executives are sitting around the boardroom one day, thinking talking about their rate plans and advanced data services when suddenly someone asks, "What about the toll-free stuff?" The room falls silent and the executives eye each other. Quickly, they elect someone who didn't make it to the meeting (or worse yet, they name the poor fool who asked the question) as the new Director of RespOrg and then call the meeting complete before breaking for lunch. I can't say whether or not this scenario has actually taken place, but I can tell you that the infrastructure and flexibility you find in your carrier's other departments is generally less robust when you get to the RespOrg department.

Escalation is your best friend when you have a problem with the RespOrg department. That is, if you don't get resolution to a problem, you take your concern to a higher power. And on, and on, and on, until you get the answers you need. As you push up the chain of command, you will eventually find someone who can either solve your problem or who knows someone who can solve your problem. Be mindful when you escalate your complaint that some people respond better to kindness, and some only respond to increasing levels of pressure. You may need these bridges into the RespOrg department again some day, so it is best not to burn them.

Following basic vanity do's and don'ts

If you are interested in a vanity number that spells something, here are some helpful do's and don'ts for finding an open number:

- **Don't try to fill all seven digits of the number with your keyword.** You have a better chance of finding a number if you have at least two digits left as unrelated numbers. Instead of looking for PRODUCE, try a search for *corn, beets*, or any four- or five-letter word that means the same thing. There is only one toll-free number in each prefix that spells PRODUCE, and the likelihood is that someone already reserved it.

  Don't lose sight of the big picture. You can liken the craze for vanity numbers to the frenzy for distinctive *URLs* (Uniform Resource Locators) during the dot-com boom of the 1990s. Companies dedicated entire departments to finding catchy Web addresses that weren't already in use. Maybe if they'd spent as much time on making their businesses viable, they would have survived. Keep this little anecdote in mind if you have your heart set on a vanity number. People will call you (no matter what your number) if you have something for sale that they want to buy.

- **Do convert your letters to numbers before you send the request.** If you're looking for the number that corresponds with PRODUCE, and you don't care if it is an 800, 888, 877, or 866 number, you should send of the request as 8XX-776-3823, not as 8XX-PRODUCE. There is always the potential for human error when converting letters to numbers from a telephone keypad. Don't leave it to your carrier to make a mistake and end up giving you a toll-free number that spells PRODUCK. If you convert the letters to numbers, and check your work before you send the request, you reduce the chance for errors.

- **Do provide concise search criteria.** Don't send a vanity request, with no number listed, and just a note at the bottom of the form asking the carrier to "please reserve a number that is easy to remember." This statement is very problematic. The main issue with this request is that the national SMS database can't be searched for 8XX-*something-easy-to-remember.* Even if the carrier finds a number that is sing-songy or repetitive, the number might be undesirable for other reasons. For example, your carrier might think that the number 888-444-8244 is a great number and easy to remember. That might be true, but if you are using the number to market to the Japanese community, where the number 4 represents death, it might not be well received. Similarly, a phone number that starts with 666 might be easy to remember, but it's not a good number for a church's 24-hour confession line.
- **Do list your search as many ways possible.** If you are looking for a vanity number with the word *corn* in it, send the request as follows:
  - 8XX-267-6XXX
  - 8XX-X26-76XX
  - 8XX-XX2-676X
  - 8XX-XXX-2676

  These options cover every possible configuration a toll-free number can have with the word *corn* in it. You might receive a list of three or four reservations for your number. After you review them, you can activate the best one. This query method allows your carrier to execute a concise search for your number, and gives you the greatest potential for success.

After you have reserved your toll-free number, *do not* publish it until the number is active *and you have successfully completed a call on it!* I cannot stress this point enough. Everyone who handles your toll-free reservation request has the potential to make a typo. If you haven't released the number to the public when you discover it isn't working because someone transposed a few digits, it isn't a big problem. If you have the number printed on 50,000 calling cards and sitting in gas stations from Miami to Manhattan when you discover the number was transposed, it's a catastrophe.

Migrating a Toll-Free Number

Most companies do not order new toll-free numbers very often. If you are in an established company, you might interact with your toll-free numbers only when you move them from one carrier to another for one of the following reasons:

- To get a more attractive per-minute rate
- To flee from bad service from your current carrier
- To gain access to more available features

The migration process is the same, regardless of the carrier you're leaving behind, the carrier you're moving to, and whether the number is dedicated or switched. It's also a potentially frustrating and confusing time for everyone, and not without its dangers. The following sections lead you through the migration process.

Filling out the RespOrg LOA

The *RespOrg LOA* (Letter of Authorization) document, generally referred to as just a RespOrg, is the most important piece of paperwork you need when you migrate a toll-free number from one carrier to another. This form identifies the toll-free number and all the pertinent information your new carrier needs to request the number be released from your current carrier. Figure 9-1 shows a standard RespOrg form.

The RespOrg LOA form is very important; fill it out completely. There are key areas of the form that need to be accurate; otherwise, you risk having your release request rejected. Every RespOrg LOA form has the same basic sections, although some carriers don't require you to list the ring-to number, area of service, or if the number is switched or dedicated. The sections that appear on every RespOrg LOA are as follows:

- **Toll-free number:** Simply list the number you want to move to your new carrier. Be sure to use the correct prefix, such as 800, 888, 877, or 866.
- **Current RespOrg ID:** The RespOrg LOA document almost always has an ID number on it, but it's really not that important. The first thing your new carrier will do when it receives the document is validate the current RespOrg ID through the SMS database. If you know the RespOrg ID for your current carrier, great; if you don't, you can leave this one blank.
- **Legal mumbo-jumbo:** Every RespOrg form has its required quantity of legal gibberish to protect your carrier if you try to take someone else's toll-free number or if the migration goes horribly wrong and the universe implodes.

ORDER INFORMATION
Type of Change: ☐New ☐Add to Account ☐Admin. Change ☐Partial Disconnect

Letter of Agency

The undersigned hereby authorizes Bogus Telecom ("Bogus") to act as the Responsible Organization ("RESPORG") for the following toll-free (8XX) numbers. The undersigned understands that this authorization is in accordance with all applicable Bogus Telecom state and federal tariffs and any accompanying terms and conditions therein.

Toll-Free Number	Current RESPORG ID	New RESPORG ID	Ring-to Number	Area of Service	Switched/ Dedicated
888-555-1234	MC101	BOG01	305-555-1234	☐48 ☒50 ☐Canada ☐VI/PR ☐International	☒ Swi ☐ Ded
				☐48 ☐50 ☐Canada ☐VI/PR ☐International	☐ Swi ☐ Ded
				☐48 ☐50 ☐Canada ☐VI/PR ☐International	☐ Swi ☐ Ded
				☐48 ☐50 ☐Canada ☐VI/PR ☐International	☐ Swi ☐ Ded

By signing this form, the undersigned also acknowledges that if this is a new toll-free (8XX) number, this toll-free (8XX) number will not be assigned to the undersigned until the toll-free (8XX) number is actually ringing to the ring-to number listed above. The undersigned further represents, warrants, and agrees to indemnify, defend, and hold Bogus Telecom harmless from any damages that may arise from this new toll-free (8XX) number not being available to the undersigned.

Understood and Agreed:

Signature *(required)* Date

Company Name (as listed on current Billing Invoice):		
Contact:		
Title:		
Service Address:		
City:	State:	Zip Code:
Phone Number:	Fax Number:	

Figure 9-1: The standard RespOrg LOA document used for migrating toll-free numbers.

- **Signature and date:** This is probably the most important section of the document. The name on the signature must match the contact name on the invoice from your current carrier. If John Smith signs the RespOrg and Mary Jones is the person your old carrier deals with, the release request might be rejected.

  The date is also very important, because RespOrgs are good for only 30 days from the date they are signed.

- **Company name:** This is the company name as it appears on the invoice you currently receive from your carrier. If you have changed your name from John's Produce to Anderson Family Co-op, yet your carrier still invoices you as John's Produce, you need to fill out the RespOrg as John's Produce. It doesn't matter that your company's name has legally changed. Until your name has changed in the eyes of your carrier, the carrier will reject any release request from a company name other than John's Produce.

- **Company address:** If your company has moved yet your current carrier is still sending your invoice to the old address, you need to list the old address on your RespOrg. Whatever you send on the form must exactly match what your carrier has listed for your billing address, even if the information is wrong. Consistency is more important than accuracy.

Scheduling your migrations

It's very important to schedule the migration of your dedicated toll-free numbers. The migration process is usually seamless for regular toll-free numbers that don't terminate to a dedicated circuit. As your new carrier updates the national SMS database, all calls from that moment on are routed to the new carrier to terminate to your phone number. You won't even know that anything has changed. When migrating toll-free service for a dedicated toll-free number, you have to be aware that the dedicated circuit must be installed before calls can be directed to the number. You don't want to put the cart before the horse.

If you begin the migration process too late, the numbers can be rejected and not be available when your circuit is activated. The industry standard for migrating toll-free numbers is seven to ten days. This means that from the day your RespOrg LOA is sent to your old carrier, the carrier has a maximum of ten days to either release or reject the migration request. You may want to pad that time frame by a day or two, if you have less confidence in your new carrier and want to give the carrier some time to process the request before it is sent to your old carrier.

Submit RespOrg forms to your carrier on Wednesday or Thursday of the week if you have any anxiety about your numbers being released. If you submit the forms on Wednesday or Thursday, the seven to ten day completion time happens between Monday and Thursday. There is nothing worse than finding a problem at 5 p.m. on Friday when the carrier's support staff has already gone home for the weekend. If there's any potential for disaster with your toll-free number migration, you want to schedule it when there is the greatest number of support staff at your immediate disposal.

Migrating too early

Don't migrate toll-free numbers before they can be used! If you migrate toll-free numbers when you begin your order for a dedicated circuit, they could be released to you in seven days and then sit idle. While you wait for your dedicated circuit to be completed (which might take over a month), the traffic for your toll-free numbers is still being sent over your old carrier. This is potentially dangerous, because carriers commonly check their networks once a month and disconnect any toll-free numbers for which they are receiving traffic, but don't have RespOrg control of. If your dedicated circuit installation is delayed, and you wake up one morning to find that your old carrier has canceled all of your toll-free numbers, you can ask your old carrier to allow you to send traffic for another few days, or you can ask your new carrier to point the toll-free numbers to a regular phone line until your dedicated circuits are tested and ready to go.

Migrating too late

If you attempt to migrate your numbers too late, your dedicated circuit may be ready for activation, but you can't install it until your toll-free numbers are released. If migrating the numbers slipped your mind, you have to wait an additional seven to ten days before you can activate both your dedicated circuit and your toll-free numbers. If the initial migration attempt fails, you will need that time to resolve the rejection and resubmit the RespOrg.

Every time you submit a release request for a toll-free number, the clock starts all over again. Even if you only have three days before your old carrier is supposed to disconnect your dedicated circuit, releasing the numbers takes a week or more.

Understanding the migration process

In order to consider the RespOrg document in context, you need to see how it flows through the telecom universe to make everything happen. Say you're currently with MCI for your long distance, and you are moving your toll-free numbers to your new carrier, AT&T.

- AT&T sends you a RespOrg LOA with its company logo on it and legal mumbo-jumbo that says you are moving to its RespOrg code ATX01. You fill out the RespOrg form, sign it, date it, and fax it back to AT&T.
- AT&T receives the RespOrg form, logs it into the system, and then faxes it to the RespOrg department at MCI.
- MCI receives the RespOrg form and checks out all the information to validate the toll-free number, the company name, address, contact name, and signature. The good people at MCI also ensure that the date on the form is not more than 30 days old. If there are no problems, MCI releases the number to AT&T.
- AT&T now has RespOrg control of the number in the national SMS database, but all the calls to your toll-free number still route through MCI. The calls won't be sent through the AT&T network until AT&T updates the national SMS database to direct all calls for your toll-free numbers onto its network.
- AT&T updates the national SMS database. All calls to the toll-free number now route to the AT&T network and are directed to the phone line at your office that you've specified should receive the calls. AT&T may associate the toll-free number to a regular phone line at your office, or to a dedicated circuit you have ordered for the sole purpose of receiving these calls.

Before you migrate a toll-free number to a new carrier, call your old carrier's customer service department and explain that you're leaving. The carrier will respect the heads up and the rep with whom you speak can let you know if anything could hinder the release of the number. A good relationship as you are leaving is helpful for everyone, because the world of telecom is very small. In a few years, you might be using your old carrier again.

Handling Toll-Free Rejection

Just because you submit a migration request doesn't mean you're guaranteed that the number will be released. Carriers have your best interest in mind when they scrutinize your RespOrg. For example, a company trying to migrate a toll-free number to a new carrier might have written down your toll-free number in error. It's your carrier's responsibility to protect your toll-free number from being migrated against your will. Carriers also use the power to reject toll-free migrations to hold your numbers hostage if you have a large unpaid balance.

A rejected toll-free migration forces you to begin the entire process over again from scratch. When you receive a rejection, you must take corrective action to resolve the issue, resubmit all the required documentation, and wait another seven to ten days for the number to be released or rejected again. The following sections list the common reasons for rejection, along with some resolutions.

All-data mismatch

An all-data mismatch occurs when the company name, address, and contact person listed on the RespOrg don't match what the releasing carrier has on file for the toll-free number. An all-data mismatch rejection is generally caused by writing down the toll-free number incorrectly; it is much less common for someone to list nothing but incorrect customer information.

To fix the problem, check the toll-free number listed and match the company name and address with what you have on the invoice from your carrier. After you have confirmed all the information, refax the RespOrg LOA to your new long-distance carrier along with an invoice from your current carrier that shows the company billing name (with address) and the toll-free numbers you are migrating (this might require sending two pages of your current invoice).

RespOrg illegible

When a RespOrg is illegible, you can usually blame the quality of the fax, which degrades the more often it's faxed from place to place (to your new carrier, and again to your current carrier).

You only need to fax your documents to your new carrier. It is the new carrier's job to negotiate the release of your toll-free numbers from your old carrier. If any of the information on your RespOrg LOA is difficult to read when your new carrier receives it, the form might be completely illegible by the time it arrives at the final destination. If your toll-free numbers are rejected because the RespOrg is illegible, complete the document again and print out all the information. When you have a beautiful copy, fax it, using the highest quality settings on your fax machine.

LOA is expired

If your RespOrg LOA was signed over 30 days before the date it is submitted, you have to fill out a new form with a fresh signature and date. If you submit a RespOrg form that is almost out of date, your current carrier might wait for your LOA to expire before rejecting it. For example, if it arrives with 4 days before it expires, and the carrier has 5 days to respond, the carrier can let the LOA sit for an additional 24 hours and then kick it back to you. To avoid this problem, always submit a RespOrg LOA that's fewer than five days old.

Name/number or address mismatch

A name/number or address mismatch indicates that the carrier is rejecting your RespOrg because the name and address don't match, even though your carrier has confirmed that the toll-free number belongs to your company. How's that for frustrating?

To correct this problem, check the address and contact person on the invoice for the toll-free number and resubmit the RespOrg with a copy of your invoice to validate what you have listed. You will probably need to send two pages of your invoice to prove the address and contact name listed, as well as the fact that the toll-free number is on that invoice.

Forging legal documents is a bad thing and can lead to time in jail. I haven't heard of anyone doing hard time from forging a name or changing a date on a RespOrg LOA, but just because nobody has been convicted doesn't mean it can't happen. Some companies forge the signature of a person who left the company but who is listed as a contact on the carrier invoice. I suppose they think that's easier than changing the contact name with the current carrier before trying to have the numbers released again. Forging isn't a good idea, but some people do find it effective as a short-term solution.

Number under contract

If you have signed a term agreement with your carrier, the carrier can reject any attempt to migrate a toll-free number until the term is complete.

Your contract might state that you will have service with the carrier for a specific duration of time, and it may state a revenue commitment, but if nothing in the contract links your toll-free numbers to the contract, call your current carrier to negotiate the release of your toll-free numbers. If, even after the release of the toll-free numbers, your company will still have enough usage on the carrier's network to fulfill your contract requirements, the carrier probably won't hold on to your toll-free numbers. After your carrier gives you the green light to migrate your toll-free numbers, you have to resubmit the RespOrg to the new carrier.

Unsatisfactory business relationship

A rejection based on *unsatisfactory business relationship (UBR)* means only one thing: your carrier thinks you owe money. The only way to work through a UBR rejection is with a few phone calls and a checkbook. If you have withheld payment because of a billing dispute with your carrier, you might be in for a bitter battle over your toll-free numbers.

Keep a clear head when you talk to your carrier, because if you tell the rep that you won't pay your invoice, you might find that the next thing you know your toll-free numbers have been immediately taken down. Diplomacy is crucial to settling a UBR. If all else fails, you may need to resort to diplomacy by other means, as discussed in the following sections that cover NASCing.

NASCing (Migration by Other Means)

Carl Von Clausewitz once referred to war as "the conduct of diplomacy by other means." Along the same lines, NASCing can be considered toll-free migration by *other means* because it bypasses the normal process and forcibly extracts the number from the carrier. The term *NASC* doesn't have to do with the National Association for Stock Car Auto Racing. It is derived from the 800 *Number Administration and Service Center (NASC),* which negotiates the release of a toll-free number. In telecom parlance, the process of breaking your bond with your current toll-free carrier is called *NASCing (nask-ing).* NASCing comes in handy in any of the following situations:

- **You have a small quantity of toll-free numbers:** There is a one-time fee of about $40 per toll-free number to NASC. If you have 500 toll-free numbers that you want to NASC, changing carriers had better be worth $20,000 to you.
- **Your initial toll-free migration has been rejected for any reason:** You can only NASC a toll-free number if the migration attempt has been rejected. If the migration attempt is simply taking too much time, your carrier will require you to wait for the number to be rejected before you can NASC it.
- **Your numbers are in jeopardy of being disconnected:** If you are in a billing dispute with your carrier and the carrier rejects your RespOrg LOA request because you have an unsatisfactory business relationship (UBR), you may be in a situation where disconnection is imminent. If you are in an intractable mess, your only choice might be to bite the bullet, pay the NASC fees, and take your numbers. Consider this option if your carrier's collections department is playing hardball and isn't validating with the billing department the unresolved disputes.

If your carrier blocks all your toll-free numbers for nonpay, you only have two options to choose. You can suck it up and pay the balance, or you can suck it up and pay the fee to NASC for all of your toll-free numbers. In any case, you're not getting out of this situation without forking over some money, so do your company a favor and choose the option that costs the least amount of money, even if that means paying money you don't think you really should have to pay. In other words, you may have to swallow your pride. On the other hand, if phones are integral to your profits, there is one benefit to spending more money to NASC your toll-free numbers than you would pay to your current carrier to end a billing dispute; if the numbers have been blocked and you need them in working order immediately, the turnaround time from when you submit the NASC request is generally 24 hours or less.

Understanding that NASCing is a temporary solution

Just as you can NASC a toll-free number from a carrier, a carrier can NASC that number right back again. And so, like the products of any unhappy marriage, your toll-free numbers are the victims of your custody battle.

Let me be clear: As long as the carrier submits the required paperwork, you and the carrier can play this NASC game indefinitely. The ownership of toll-free numbers doesn't usually degenerate to NASCing wars, but it isn't unheard of if the amount of money at stake is large enough.

Check all of your documents several times before you submit a toll-free number to NASC. If you mistakenly NASC someone else's number, you can expect bad things to happen. At a very minimum, you will get a screaming call from the number's rightful owner. If you take down the main order number of a huge company, you might hear terms like *civil suit* and *punitive damages* bandied about in the very near future. Carriers that NASC numbers with reckless abandon have been fined millions of dollars and required to follow stricter quality-assurance policies.

Following best NASCing practices

The industry standard timeline for NASCing is 24 hours if the toll-free number is blocked and nonoperational, and the standard 7 to 10 days if it's active. Aside from the timeline based on the number's disposition, the paperwork you must submit to make the NASC happen is the same (that is, you fill out a RespOrg LOA). To NASC a number, your initial migration has to have been rejected, and your new long-distance carrier will most likely ask you to submit the following:

- A new RespOrg LOA form filled out completely, with a current signature and date
- Pages of the most recent invoice for the toll-free number that validate both the number and the company information

After your new long-distance carrier receives your fax of the documents and you explain the urgency of the NASC (is your number active or down?), the carrier validates that the number has a current rejection file and begins the NASCing process.

Your new long-distance carrier sometimes calls the number before beginning the process in order to validate the company listed on the RespOrg. This is helpful only if the number you are NASCing isn't down or a fax line, because you can't validate the company on the end from either call treatment. If you are trying to NASC a number that belongs to a subsidiary of your company, or that's used for a specific project and you fear that the call won't be answered with a clear reference to your company's name, you should alert your new long-distance carrier. If your new long-distance carrier makes a test call to the number and the name of the company answering the phone doesn't match, it may not execute the NASC until you can explain the disparity. You can bypass this potential problem by reviewing your invoice with your carrier.

If your carrier is required to make the test call before NASCing because of the results of a federal lawsuit (possibly because of reckless NASCing in its recent past), the carrier might resist going to the extra work. If you need to, call your new long-distance carrier and set up a conference call with someone in your office who answers toll-free calls. After everyone is on the line, tell your carrier to call the number and have your co-worker answer it by identifying your company by name ("Thank you for calling Acme. This is Mark. Can I help you?"). Making one clean call takes only five seconds, and then everyone will be happy. Let the NASC begin.

Ordering Switched Toll-Free Numbers

All the information in this chapter up to this point applies to all toll-free numbers, whether they ring to a switched phone number or into a dedicated circuit. Now I focus on switched toll-free service. Ordering a switched toll-free number without any advanced features is so basic that everything you need to include is usually added onto the standard RespOrg LOA document (refer to Figure 9-1). The only information your carrier needs to activate a switched toll-free number is the toll-free number, the ring-to number, and the area of coverage:

- **The *ring-to number* receives the calls from your toll-free number:** If you have five phone numbers that roll over to each other (what the telecom world refers to as a *hunt group,* because the calls hunt from one line to the next), you point your toll-free number to that first phone line so you can take calls on all five lines. The first number in the hunt group is your ring-to number.
- **The area of coverage specifies the geographic area that your toll-free number serves:** You can have one default area of coverage, or these options:
  - U.S. 48 states only
  - U.S. 48 states and Canada
  - U.S. 48 states, Puerto Rico, and U.S. Virgin Islands
  - U.S. 48 states, Canada, Puerto Rico, and U.S. Virgin Islands
  - U.S. 50 states only
  - U.S. 50 states and Canada
  - U.S. 50 states, Puerto Rico, and U.S. Virgin Islands
  - U.S. 50 states, Canada, Puerto Rico, and U.S. Virgin Islands

Rates for toll-free calls from any area outside the 48 contiguous states are typically higher than domestic rates. Check the costs to determine whether it's worth it to allow them access. If you have questions about the pricing for toll-free numbers, please visit the section about the costs of enhanced toll-free service in Chapter 5. While you're there, you can read up on the enhanced toll-free features that might be useful for your business as well.

After you reserve a new toll-free number or successfully migrate an existing toll-free number to a new carrier, activating the number takes about 24 hours. The actual process of updating the national database to point the traffic from the toll-free number to the new carrier takes about 15 minutes, but your carrier is busy processing hundreds (maybe thousands) of orders, so you can wait a day, can't you?

Ordering Dedicated Toll-Free Numbers

Dedicated toll-free numbers are inherently more complex because the toll-free number isn't simply being pointed to a phone number and sent off to your local carrier to complete. The number is instead being designed to follow a predetermined route within your carrier, terminating at your specific dedicated trunk group. That's what the *dedicated* in *dedicated toll-free* service means.

Dedicated toll-free ordering timelines

There aren't any industry standards for activating dedicated toll-free numbers, and the time frames vary greatly from carrier to carrier. One carrier might activate a new dedicated toll-free number in 24 hours, whereas another carrier takes 10 to 15 business days to do the same thing.

You need to call your carrier to find out its standard timeline so that you can plan accordingly. Be prepared to see longer timelines for toll-free numbers that must be migrated from another carrier or for change orders.

When you have an order for more than 100 toll-free numbers, you can usually go through a special bulk-load or bulk-ordering process that shortens the overall activation time. Instead of having to schedule time for someone to design all the toll-free numbers individually, *bulk loading* can cut the time in half. Your carrier might have a minimum quantity of toll-free numbers before it allows bulk-load activation; check into this option if this factor might affect your decision about how many lines to order.

Giving general information about the order

Before you can add any enhanced features to your dedicated toll-free numbers, you need to fill out a standard order form for dedicated toll-free service. If you are unfamiliar with the standard features on dedicated toll-free numbers, check out Chapter 5, where I cover them in depth. The order form for a basic dedicated toll-free number asks you for accounting information, such as your company's name, account number, and contact information, as well as the following technical info:

- **The toll-free number:** You gotta know that much!
- **The required coverage area:** Specify whether you want to receive calls from mainland U.S. states, Canada, Alaska, Hawaii, Guam, the U.S. Virgin Islands, or any combination of these geographic locations.
- **The dedicated trunk group to receive the toll-free number:** Every carrier has its own requirements and terminology for toll-free numbers. It's more common for carriers to require you to list the trunk group that they should terminate into than to list the circuit ID. Trunk groups are configured to designate how your carrier sends calls, and a single trunk group may include multiple circuits or only a few channels on a single T-1 circuit. Check out Chapter 8 for more information about trunk groups.

Ordering DNIS digits

Most people in telecom use the last four digits of the toll-free number as their *DNIS digits.* If you have several vanity numbers whereby the last four digits of the toll-free are the same and you need to differentiate where the numbers are directed by your phone system, you can't use this method. Instead, you can assign random DNIS digits to your toll-free numbers.

A matter of routing

The reason for this huge disparity in time — from 1 to 15 days — is all because of how your carrier builds your toll-free numbers. Most carriers that take more than 24 hours to activate a dedicated toll-free number build individual route paths for them. This is a manual process whereby every toll-free number is keyed in by a technician.

If your carrier builds a single routing pattern for all toll-free numbers on your circuit, it can probably turn up new numbers in 24 hours. In this case, all the technicians have to do is link the new toll-free number to that generic route. Check with your carrier to determine which procedure it uses, because each option has limitations.

Waiting 10 to 15 days is certainly not desirable. However, if your carrier builds a boiler-plate route for your company, there might be DNIS limitations that prevent you from using the last four digits of your toll-free number as the DNIS on your phone system. This problem is common when the last four digits of the toll-free number start with a 0, because some phone systems can't process a DNIS with a leading 0. Because you now have a toll-free number that doesn't adhere to the profile used for the rest of your toll-free numbers, you need to speak to your long-distance carrier to find a workaround. If you have questions about DNIS, check out Chapter 5.

If you go for a unique DNIS setup, keep a log that explains which DNIS digits correspond to which toll-free numbers. It's easy to become confused if you have to troubleshoot your phone system, and a simple spreadsheet can be a lifesaver.

Configurations for DNIS can vary within your phone system, so you might end up searching for the correct DNIS option. If you have a long-distance T-1 with DNIS, and are having a difficult time finding the correct configuration in your phone system, try the option for *Local DID;* these settings are generally compatible.

Setting up ANI delivery

If you need to display the phone number of the person originating calls to your toll-free number, you need to order *ANI delivery.* Some phone systems can be configured to capture this information and list where your calls are coming from on a specific toll-free number. If you launch a marketing campaign in Chicago, you can get an idea of how effective it is by determining how many people in Chicago called in on your toll-free numbers.

Your carrier might not be able to provide ANI delivery if you also need a ten-digit DNIS. Many carrier switches have limitations on the number of digits they can input in the DNIS stream. If the switch is restricted to 20 digits and you need ANI delivery, your DNIS can't be longer than 7 digits, because the rest of the available slots are taken up with asterisks (*). The information is sent in the DNIS stream in this scenario as:

```
*(10 Digit) ANI*(7 Digit)DNIS*
```

The only time you need all ten digits for DNIS is if you literally have thousands of toll-free numbers — say you own 1-800-FLOWERS, 1-888-FLOWERS, and 1-877-FLOWERS. Because the last seven digits of the toll-free number are all the same, the easiest thing is to simply use a ten-digit DNIS. A ten-digit DNIS prevents you from maintaining a spreadsheet of all your toll-free numbers and the unique DNIS digits for each of them.

Setting up ANI Infodigits

If your business plan requires you to identify the type of phone originating calls to your toll-free number, you must use *ANI Infodigits (ANI II).* ANI II is generally used for companies that provide calling-card service and must account for surcharges applied to calls made from pay phones (infodigits 27, 70, and sometimes 07).

What DNIS actually looks like

Dialed Number Identification Service, or DNIS, involves a challenge and a password between your long-distance carrier and your PBX. The whole process is accomplished by the use of dial tone and DTMF digits (the standard tones you hear when you press the button on a touch-tone phone). Your phone system is alerted to the fact that a toll-free call is coming in and connects the call. Your phone system then sends a signal back to your carrier, which, for all intents and purposes, sounds just like a dial tone. Your long-distance carrier receives the dial tone and understands that you want the carrier to send the DNIS information, at which time the carrier sends you this:

```
**DNIS*
```

Each asterisk is actually the same touch-tone sound that you hear when you press the * key on your phone. The DNIS is made up of the two to ten digits you ordered to be associated with that specific toll-free number. This configuration isn't an absolute in telecom, because all carriers are different, but it's the most common you will encounter.

ANI Infodigits aren't available on all circuit protocols. Unless you are using ISDN or Feature Group D, you cannot use ANI Infodigits. Please review the sections in Chapter 8 that cover protocols.

Overflow routing to another dedicated circuit

Dedicated toll-free order forms allow you to list the primary trunk group that receives the traffic for the toll-free number, as well as a secondary trunk group. If you want the calls to roll over to more than one more trunk group, you can squeeze all the trunk groups in sequence onto the form, or simply list the overflow routing sequence in the notes section at the bottom of the form.

Overflow routing to a switched phone line

It's very common to set up overflow routing to a switched phone line, and this configuration is always the final route (the end of the line, so to speak) for a dedicated toll-free number. After a dedicated toll-free number overflows to a switched phone line, the network doesn't route the call any farther. The only way to give the toll-free number more life when it hits a switched phone

Understanding trunk group level features

DNIS, ANI, and ANI Infodigits are generally built at a trunk group level. If you set up your trunk group to use the last four digits of your toll-free numbers for DNIS, you can't request to have a toll-free number pointed there that uses a seven-digit DNIS. This rule is the same if your trunk group is set up to deliver ANI or ANI Infodigits, but for a select group of the toll-free numbers, you don't need the additional information you get from ANI or ANI Infodigits. You must either change the toll-free numbers so that they match the trunk group or build a new trunk group for these special cases.

line is to point it to the first number in a multiple-line hunt group. See the section "Ordering Switched Toll-Free Numbers," earlier in this chapter, for more on hunt groups.

Identifying new or migrated numbers

Carriers have different requirements for orders that are new toll-free numbers that are reserved, as opposed to toll-free numbers that are being migrated from another carrier. Carriers can process new toll-free numbers at any time without any concern. If the new toll-free numbers are activated and begin ringing into your phone system at 2:00 a.m., there is no real downside.

On the other hand, if you're moving all your toll-free numbers from another carrier to a dedicated circuit that isn't yet installed, pointing the numbers to the new (inactive) trunk group will effectively bring down your phone numbers and have them ring to busy signals until you fix the problem.

Using a Hot Cut to Activate Your Dedicated Toll-Free Numbers

You can choose to let your carrier activate your toll-free numbers at will, or you might need to have it done in a very controlled manner. If you are ordering a handful of new toll-free numbers for your circuit and you haven't published them, there isn't much threat of disaster if they aren't working right away. If the numbers are failing, you can always troubleshoot them in a relatively calm environment.

On the other hand, your business may depend on toll-free service. If you have an existing toll-free number that handles 90 percent of your inbound orders every day, having the system down for even a minute is simply not an option. You certainly don't have the luxury of waiting for someone to tell you that your toll-free service isn't working, tracking down the carrier's number, opening a trouble ticket, and then waiting for two to four hours for a call back (all while your boss is screaming at you to get it fixed).

So what *can* you do? Glad you asked! A *hot cut* is a scheduled activation whereby your carrier walks you through the activation process via conference call. A hot cut allows you to identify any problems immediately, and quickly correct them while you have a technician on the line. If your carrier cannot fix the problem immediately, let the troubleshooting begin.

Making sure the right people are invited to the hot cut party

You have to determine who should participate in the hot cut phone call and ensure that a conference bridge is available if you need one. The *conference bridge* enables many people to dial into a system where they can be placed on one large conference call. This is ideal for hot cuts because you need to have several people on the call who are geographically separated. In addition to the technician at your carrier, you may consider inviting the following individuals:

- **The provisioner at the carrier:** If you have requested any special routing features on your toll-free numbers, you might want to include your provisioner, who is the individual at your long-distance carrier who receives and processes your orders, in the call as well. In the event that there has been a miscommunication and your special features have not been added, the technician activating your number might be able to build them at the time of the hot cut. Your long-distance carrier will require you to send an order through the provisioning system after the fact, to complete the paper trail. When you include the provisioner, that person can validate anything that might not have translated through the provisioning system and place the backup orders if something was missed.
- **The hardware vendor:** If you are migrating existing, active toll-free numbers to a new phone system, your hardware vendor has to program each toll-free number into your new PBX, Key System, or whatever constitutes your new phone system, before the toll-free numbers will work. The vendor needs to build the routing plans on the DNIS (if you are using it) for each toll-free number, as well as a default path in case you are sent a call with an unknown DNIS. If you don't have a default path, and the DNIS isn't set up for a number, your phone system will reject calls. Similarly, if the DNIS stream isn't sent completely, your phone system will reject calls. If

you have existing toll-free numbers from the carrier working on your phone system and your hardware is already set up to receive the new number, you only need yourself and the provisioner and the carrier's technician on the hot cut.

Accomplishing the perfect hot cut (Timing is everything)

The key to success in telecom is planning. If you are activating new circuits, here's the perfect timeline.

- **One week or so after you submit the order for your dedicated circuit:** Send the RespOrgs for your dedicated toll-free numbers.
- **After you receive confirmation that the circuit order is flowing through the system:** Send the order for your dedicated toll-free numbers and the RespOrgs.
- **The day you send the RespOrg LOA paperwork:** Schedule a follow-up call eight days later. By then, you should know the status of the RespOrg and can handle a rejection or celebrate a release with plenty of time before your circuit is ready.
- **About two weeks before you expect to activate your circuits:** Issue the disconnect order with your old carrier and request the circuits be taken down in 30 days. This request gives you about two weeks of overlap just in case an unforeseen problem crops up.

  It's also very helpful to remain on friendly terms with your old carrier, just in case you need to make an 11th-hour call to extend the due date of your circuits.
- **Two or three days prior to activating your new circuits:** Call your new long-distance carrier and confirm the order details. Confirm the quantity of toll-free numbers pending, as well as the RespOrg of all the numbers. If any numbers are still not with new long-distance carrier, you might still have time to push them through the system.
- **On the day of the hot cut:** Have everyone at your disposal. Be sure the new long-distance carrier's installation technician is on the line, along with your hardware vendor. You also want to have the person who takes your orders for the new long-distance carrier available for verification of orders or if you need to push a straggling order through. Refer to your escalation list so you know who should be present if the person who took your order is not in the office for the hot cut. As long as your new carrier has RespOrg of your numbers, and you have competent technicians, both at the carrier and hardware levels, there isn't much that can stop you from a successful activation.

Resolving Activation Issues on Migrated Toll-Free Numbers

In a perfect world, there are no problems. All toll-free numbers are released to the new carrier on time and are seamlessly transitioned onto your new circuits. In reality, this isn't the case. Toll-free hot cuts fall into four basic scenarios; the scenarios range in type from the highly desirable to varying degrees of undesirable.

Before you can fully understand the complexities of the hot cut, you must know two important things.

- **Which carrier has RespOrg ownership of the toll-free number:** This carrier has the power to manipulate the toll-free number in the national SMS database, and can change the long-distance carrier that receives the calls for the toll-free number, as well as the location into which the toll-free number rings. After your old carrier releases RespOrg ownership of the number, the old carrier can no longer update the national SMS database to change where calls for the number are sent.

  This means that the old carrier hands off the power over your toll-free number while all of your traffic is still running over its network.

- **What carrier is actually handling the traffic and completing the calls for the toll-free number:** The traffic for your toll-free number continues to run over your old carrier's network until either your new carrier updates the SMS database to send the calls to the new network, or your old carrier blocks your toll-free numbers on its network.

  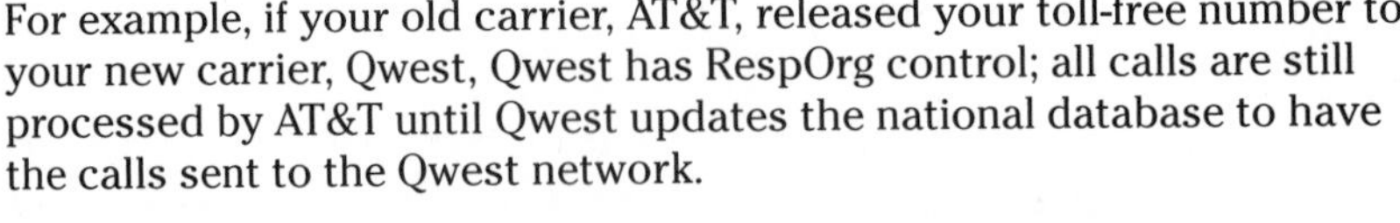

  For example, if your old carrier, AT&T, released your toll-free number to your new carrier, Qwest, Qwest has RespOrg control; all calls are still processed by AT&T until Qwest updates the national database to have the calls sent to the Qwest network.

Your new carrier should have a handful of toll-free numbers that are used for testing the inbound routing on dedicated circuits. Have your new carrier set up one of these numbers exactly as if it were your toll-free number. That way, technicians can validate the routing and configuration before they close out the order and update the national SMS database. If, for example, your main toll-free number has ANI delivery and the four-digit DNIS of 8899, ask the carrier to replicate that setting so that some calls are sent to the circuit. If your phone system doesn't respond to the calls, or if the calls simply fail, you can use the test toll-free number to work out the kinks. Your new carrier should update the SMS database only after all the issues on the test toll-free number are resolved. You should make more test calls to your numbers after the SMS database is updated, too. When your calls complete over the network and all the test calls are good, you are done.

Resolving Common Dedicated Toll-Free Migration Scenarios

It's much easier to understand the possible pitfalls of dedicated toll-free numbers if you consider a real-world scenario. For this example, your fictitious company's current long-distance carrier is AT&T, but it's moving all of its services to Qwest.

You have

- Completed the dedicated order forms with Qwest for the T-1s that will replace your AT&T circuits
- Filled out the RespOrg LOAs to move all of your toll-free numbers
- Noted that all of your toll-free numbers must be activated in conjunction with your new Qwest circuits
- Requested a hot cut for activation

In the following sections, feel free to replace the main characters AT&T and Qwest your actual carriers.

With this as the starting point for your RespOrg one-act play, the following sections reveal the possible issues that could arise, as well as offer the resolutions that will make things right again. (You can substitute any two carriers for this example, because the scenarios could just as easily happen between any two other carriers.)

Activation without a hot cut

In this scenario, your toll-free numbers are released from AT&T to Qwest without a problem, and Qwest now has RespOrg and activates the toll-free numbers to your circuits immediately. Now your circuits won't be active for another two to three weeks. All calls to your toll-free numbers fail to a fast busy signal.

Now you need to fix the problem, but you don't know who to call. Don't panic; you can resolve this problem in short order. Your first call is into the person at Qwest handling the order for your toll-free numbers to confirm the following details:

- Who has the RespOrg
- Which carrier is receiving the traffic

When you confirm it's Qwest in both cases, you now know that Qwest jumped the gun and is routing calls to your dedicated circuits that are not yet active. AT&T can't do anything to help you because it doesn't have RespOrg control of your toll-free numbers anymore, and you need to bring your toll-free number back to life.

You have two options to resolve the situation and get you back up and running until your circuit is activated.

Solution 1: Talk to whoever's in charge of toll-free routing or SMS database updates

Have your order person at Qwest connect you to a manager or the most skilled person in the RespOrg or toll-free routing department. Don't be concerned about what titles; as long as the person you talk to is responsible for updating the SMS database, you are talking to the right person. Tell this person that you need to send the traffic back to the previous route. Because you know your old carrier is AT&T, you can have Qwest update the SMS database to route all of your calls over the AT&T network. This is the preferred solution and generally the most difficult to accomplish, because only a handful of people in the toll-free department at any carrier know how to make this change to the national SMS database.

After the national database is updated, your toll-free calls for that number go back to being handled by AT&T. Calls are routed to AT&T networks until you have Qwest update the numbers to point back to Qwest when you finally complete the testing and activation of your dedicated circuit.

Whoever holds the RespOrg control of a toll-free number can send the calls over any long distance carrier in North America, Canada, and parts of the Caribbean. If your old carrier (in this case AT&T) still has an active route built for the number, your current carrier can update the database to send calls to the old carrier, and your calls will complete just fine.

Be sure to tell the RespOrg person to send the traffic back to the previous route and not to release the RespOrg to the previous carrier. These are two completely different things. As long as Qwest simply updates the national SMS database so that your toll-free calls ride AT&T's network, Qwest retains the ability to point your calls to the Qwest network when the time comes. If Qwest misunderstands you and thinks you want to give the RespOrg control back to AT&T, everyone will become very frustrated, because you need an immediate fix, and the process of switch RespOrg ownership of a toll-free number takes seven to ten days.

It seems logical that Qwest could simply give the RespOrg back to AT&T and avoid the seven-to-ten-day process. That is a bad idea, because without a RespOrg LOA telling Qwest who to give the RespOrg to, Qwest might make a mistake.

In order to make such a move work, you would have to call AT&T, have a rep fax you a RespOrg LOA form, fill it out, fax it back to AT&T, and have AT&T send the form to Qwest again. All of this takes time that you don't have, and even if you could do it, you don't want to. The solution I've offered here is a temporary fix to make sure that you get calls (through AT&T) while the dedicated circuits (with Qwest) are being prepared. Therefore, if you were to start all over again with the LOA, as soon as the number is sent back to AT&T, your dedicated circuits will probably be ready with Qwest, and you will have to start the whole process over again to pull the numbers back to Qwest.

Solution 2: Call customer service to report trouble with the toll-free number

After you get off the phone with the person at Qwest who handles your orders, call the Qwest customer service department to report trouble with the toll-free number. When you reach a Qwest customer service rep, tell him or her what happened and that you need to have an *emergency alt route* (telecom code for emergency alternative route) *to a switched number* for your dedicated toll-free number. The rep will ask you where you want the toll-free number pointed, so you need to have a phone number in mind that you can dedicate to temporarily receiving the toll-free calls. The phone number you give can be any regular switched phone line in your office, as long as it's a working number that someone will actually answer.

If you are in transition and you don't have any active phone lines in your office, ask your carrier to send the calls to your cellphone until you have a better option. No, I'm not kidding. This isn't the best solution, but it will allow you to continue receiving calls until your dedicated circuit is running. After your T-1s are active, call the Qwest customer care department to have the calls sent back to the dedicated circuit. The installation technicians might not be able to tell that your toll-free numbers are on an emergency alt route, so you need to inform them.

Cancellation before activation

In this scenario, your toll-free numbers are released by AT&T to Qwest without any problem, and Qwest remembers that you need a hot cut. In light of your request, Qwest holds the RespOrg for the numbers but doesn't cut over the traffic, and for a while everything looks wonderful.

Then something happens. Possibly the installation of your Qwest circuits is delayed, or maybe your relationship with AT&T goes sour. Whatever the situation, AT&T blocks your toll-free numbers on their network before your circuit with Qwest is active and can receive the calls. This scenario is common when there's an issue with the new circuits that pushes out the schedule for activation on the Qwest network.

Many carriers have an automated system to clean up unwanted toll-free numbers from their network. Basically, a computer program polls every toll-free number passing traffic on the network. If the toll-free number has a different RespOrg (one that doesn't belong to the carrier or any of its subsidiaries — a situation that occurs if you've transferred the number to a new carrier), the automated system blocks the toll-free number from passing traffic over its network. All the calls will either fail to a fast busy signal or will be sent to a recording that says something like, "This number has been disconnected. Please contact our customer care department if you have any questions."

If this happens to your toll-free numbers, your first call is to the person at Qwest who handles your toll-free orders. Ask

- Who has the RespOrg control of your toll-free numbers?
- Which network is receiving the traffic at this time?

When the Qwest rep tells you Qwest owns the RespOrg, but the traffic is still running over AT&T, you know why the calls are failing. Because the national database is sending all of your calls to the AT&T network and nobody else has made any changes, the problem lies with AT&T.

You have two solution options.

Solution 1: Ask for the blocks to be removed

Ask your AT&T account rep, in your nicest voice, to remove the blocks on your toll-free numbers. If the rep is in a good mood, he or she will realize that opening your toll-free numbers will get AT&T some more revenue before you leave, as well as garnering good will.

It might take AT&T anywhere from 15 minutes to a few hours, but as long as the carrier is working with you (and not against you), the numbers will be active again soon. Chat with the rep about the progress of your new circuits, especially if you have hit substantial delays. This is your preferred solution to the problem because you maintain the status quo.

If it's going to take 30 or 60 more days before you can activate your toll-free numbers on your new circuits, you might choose to change the RespOrg ownership of the numbers back to your old carrier in the meantime. Say AT&T has an automated system that takes down your numbers every 30 days; if you have to go through a fire drill to have them reactivated once a month, it's better to simply send the numbers back and start the process over again after your new circuits are ready to test.

Solution 2: Call the toll-free order people at Qwest to have them activate your toll-free numbers.

If AT&T blocked your toll-free numbers because your relationship turned sour, you can't likely rely on AT&T to bail you out of this problem. The only lines the Qwest reps can use to send your calls is your dedicated circuit (which hasn't been installed yet), but as long as they can update the national SMS database to point all calls to the Qwest network, you are halfway home.

You might need to escalate the issue if you feel it isn't being handled with enough urgency, but in a pinch, every carrier will do what it takes to restore your service.

After the national SMS database is updated, you simply have to call the Qwest customer care department to perform an *emergency alt route* to a switched phone number. About 15 minutes after you give Qwest the number to a switched phone in your office, you should start to receive toll-free calls to the switched line. I do realize that your toll-free number might have had 24 or 48 phone lines to ring into before, but you will be able to get by until your circuit is installed and you can swing the numbers onto your new dedicated circuits. If you have four or five lines that roll over to each other, you can send your toll-free number to the first number in the group, and at least not be limited to receiving one call before people get a busy signal.

Negotiating a hot cut without RespOrg

Say you submit the RespOrg LOA documents 30 days before your circuit is ready to activate, but your toll-free numbers are still not released to your new carrier by the time the circuit is ready to activate. Qwest might have forgotten to submit the RespOrg documents to AT&T, or Qwest may have sent them but the LOA was rejected and never resolved.

This situation (essentially that your new carrier, Qwest, hasn't got official status as the RespOrg owner of your toll-free numbers) is generally revealed as you are installing the dedicated circuits. You might have spent an hour or so testing the outbound calls on your new circuits before you decide to test the toll-free numbers. As you casually ask your installation technician for Qwest to check out the toll-free numbers, she responds, "Hmmm, I don't see any setup for this circuit." Ten minutes later, you find out the location of your toll-free numbers — with AT&T. Regardless of the source of the problem, the end result is that Qwest doesn't have RespOrg of your toll-free numbers and you cannot migrate the numbers. The really painful part of the situation is that you only have seven days (or fewer) before your AT&T circuits are slated to be disconnected.

Don't panic yet. Here are some solutions if you find yourself in this situation:

Solution 1: Push out the cancellation date

Call your contacts at AT&T and push out the cancellation date of your circuits by another week or two. If the reps are nice, you can buy yourself enough time to migrate your toll-free numbers to Qwest. If the reps are not inclined to help you, opt for Solution 2.

Solution 2: Submit the migration request — yesterday

If the migration request hasn't been submitted to AT&T yet, have it submitted immediately. If the RespOrg has already been rejected once, you can address the reason for rejection and resubmit the migration request, or simply NASC the numbers. If you have a large quantity of toll-free numbers, you need to perform a little telecom triage to determine which numbers are the most important and cannot go down (as I explain earlier in this chapter, NASCing *is* rather expensive if you have many lines). Some of your numbers might have limited use, and so it might be just fine to wait for the seven to ten days, even if the lines aren't working for three of those days. Not ideal, but it'll work.

Chapter 10

Activating Your Dedicated Circuit and Toll-Free Numbers

In This Chapter

- Gathering the right people to activate your circuit
- Working through a preinstallation checklist
- Recording the installation in a journal
- Connectivity, hardware, and configuration problems
- Troubleshooting connectivity issues

If you're setting up a dedicated circuit for your toll-free numbers, the most critical time in the process is the moment you unplug your old circuits and plug in your toll-free numbers to your new carrier. During this time, you have no phone service. You won't know if your new circuits function until your carrier activates the numbers on the circuit and you successfully transmit your first call through the new network.

There are a handful of variables that can influence how long the installation takes (and, consequently, how frustrating it can be). During the installation, all focus is on the hardware at your end, the hardware at your carrier's end, and all the hardware in between. Your greatest allies (and, potentially, your greatest sources of frustration) during the installation are the technicians at your disposal. You must learn to love your hardware vendor and the installation technician at your carrier.

For basic information about your toll-free options, check out Chapter 8. For introductory information about all the stuff you need to do to order your toll-free dedicated service, move to Chapter 9. This chapter walks you through a typical installation of a dedicated circuit that can be used for your toll-free lines, or strictly for outbound calling, and ensures that you have access to all the right people. This is the chapter you should read if you're about to activate any dedicated service.

This chapter covers installing brand-new toll-free numbers on brand-new dedicated circuits with a brand-new carrier. If you're setting up a new dedicated circuit with a carrier and *migrating* existing toll-free numbers from your old carrier to the new circuit, flip back to Chapter 9 and look for information about *hot cuts.* Go ahead. I won't be mad.

Requesting a Hot Cut or Parallel Cut

A *hot cut* on a dedicated circuit entails the same type of risk as a hot cut on a dedicated toll-free number. Your active dedicated circuit is taken down and replaced with a new dedicated circuit. Until the new circuit is fully activated, you can't use any phone service that depends on the dedicated circuit. This type of a circuit installation should be conducted after the close of business or during off-peak calling times.

A parallel cut on a dedicated circuit involves the activation of a dedicated circuit that is to be used in addition to your existing service. None of your current phone service is in jeopardy of being disturbed, so this type of circuit activation can occur at any convenient time during regular business hours.

Regardless of whether you're replacing your existing dedicated circuits on a hot cut, or whether you're simply adding additional circuits on a parallel cut, the same rules apply to both.

Inviting the Right People to the Installation

There are only a handful of people you want on the call while you are activating a circuit. These people can be grouped into two categories: the people you must have and the people you may need. Of course, there's also a third category — the people you definitely do not want. Check out the following sections for more information on which people are in each group.

Getting your A-list together: Mandatory participants in your activation party

Here's a list of people who must be present when you activate your toll-free dedicated circuit:

- **Your carrier's installation technician:** Your carrier has to individually activate every DS-3, DS-1, and lower-level DS-0 on your circuit. If the technician doesn't show up, all bets are off; you can't install the circuit.
- **A technician from your hardware vendor:** You must have someone technically capable on your end to validate settings on your hardware and to perform simple testing. If you do not have a hardware vendor, hire one for this event. An installation tech that can bring your circuits up without any problem is well worth the investment.
- **You:** Someone needs to write down everything that happens during the installation. That someone is you, unless you can think of someone more trustworthy and knowledgeable than you (in that case, I still think you should be there, along with this other ace). You won't be activating any circuits. You're there to witness and record all the vital information about the birth of your beautiful new dedicated circuit. Bring a pen and paper (see "Writing an Installation Journal," later in this chapter), and record the name of the technicians present, along with the general sequence of events. Try to make your notes as specific as possible without going overboard.

  I suppose you could sit in your office instead of being present for this technical achievement, but it's really not a good idea. For one thing, if you are not involved in the installation, there may be very little motivation to successfully complete it. For another thing, you may actually be useful. And finally, you could save your company a lot of money by being there. Your hardware vendor is probably charging you by the hour, so even though he has other work to do he's not necessarily going to work at breakneck speed. Likewise, the technician at your carrier may not feel the need to push through simple issues without proper motivation. Issues that have financial implications can come up during an installation process. The only way to avoid being told that the installation went fine, and then realizing a month later (when the bill arrives) that something costly has occurred, is to be there when it happens. Say your hardware vendor changes the configuration on the five T-1s you installed from loopstart to ISDN. Although this change enables additional information to be transmitted in the call stream (which may be useful for your business), it may also cause you to incur additional charges. If you're not there to say "Heck, no," your carrier may charge you $250 for the change, $250 apiece to install the five D channel cards, and $150 per month for the cards. The installation technician can't waive those fees, and you will have a difficult time renegotiating the price with your salesperson on a feature that has already been installed.
- **Your local carrier's installation technician:** If you ordered a local loop (see Chapter 8), you're responsible for coordinating the testing between your long-distance carrier and your local carrier. If your long-distance carrier ordered the local loop, don't worry, the long-distance carrier will have already done its testing before even allowing you to schedule the installation.

Do not try to install new circuits on new hardware yourself. Unless you are comfortable enough with your phone system and can confidently configure, test, and confirm settings within it, you need to hire someone who can. Every phone system's configurations are set up a bit differently; some have *dip switches,* some require you to use a laptop to connect into them, and some can be set by prompts given on an LCD screen on the face of the unit. If you don't know what you're doing, you could to waste four or five hours figuring things out through trial and error.

Bringing in the B-list: Optional guests

Here's a list of people who you should welcome to your installation party. If these people are there, they can make things go smoothly if something goes wrong. But if they regretfully decline to attend, you shouldn't hold up the installation process on their account. Invite the following people with fingers crossed:

- **A rep from your carrier, specifically, the person who handled your order:** If there is a problem during the installation where an order was either not placed correctly or not completed correctly, you need the rep on the line to correct the issue. If the issue is small, it may be resolved during the installation and paperwork can be sent in after the fact to tidy up the accounting. If there is a large problem, your provisioner may have to take the issue to his or her manager to clear it in a timely manner. If your carrier rep isn't on the line during the installation, he or she should at least be available by phone during that time.
- **Another hardware vendor:** If your hardware vendor is technically incapable of performing the tests and setting confirmations that are required during the installation, you should consider replacing the vendor before the process even begins. If the installation is going south, a backup technician may be very useful.
- **Building management:** If the installation has problems, your hardware vendor may need to gain access to the main phone room for the building. This room is generally locked and on the first floor of the building, or possibly even in the basement. If you're activating a circuit after normal business hours, you should arrange for access to this room if necessary.

Banning participants: You definitely don't want to see these faces

Here's a list of people you definitely don't want to come to your installation party. If these people are there, they can hinder productivity, cause confusion,

Working the diplomacy angle to resolve handshaking issues

There are issues that occur, not because of a hardware fault at either end of the circuit, but actually as a result of the interaction between your carrier's hardware and your phone system. These *handshaking issues* can generally be resolved by validating all the line coding, framing, and configuration information seen by both your carrier and your hardware vendor. If that information checks out, the issue could be of a more granular nature that needs more troubleshooting.

It is at this time of mounting frustration that you must be your most diplomatic. The hardware at your carrier's end may pass self-diagnostic tests with flying colors. In fact, your phone system may also pass self-diagnostic tests, giving both your vendor and your carrier validation for feeling like it's the other guy's fault if the system fails. Technicians for both pieces of hardware can claim that whatever is causing the problem is not their issue. At this point, only their respect for you, your skills as a diplomat, and their commitment to the job (in that order) will allow you to push through to resolution.

If you are in the middle of an installation that begins to degenerate, you may have to separate your hardware vendor from the installation technician at your carrier. This may be the best way to get everyone to do what they need to do without anyone feeling threatened. Until you have been there, you won't believe how easily things can move to problem solving to a systematic game of trying to prove the issue is someone else's fault. Meanwhile, of course, the technicians ignore you, your phone system, the configuration process, and so on. On your dime.

If you are 99 percent sure you know where the problem is, there is a way to get Sam the tech to do additional testing, even if Sam's irritated and emotional because he has been attacked for the past three hours. The technique begins when you calm down and call him up, without anyone else on the line. In your most understanding voice, say, "I know the problem is not on your end, but my hardware vendor is flipping out and I just want to shut him up. Can you do this one test, just to prove, beyond the shadow of a doubt, that it is not in your network? Then I can make the vendor fix his end of this." The technician will run the test, or even replace the faulty hardware. If the problem resolves itself, then you know he found the issue and fixed it. If the problem persists, at least you have one part of the troubleshooting done. Now it's time to call the technician on the other side of the circuit.

and make things go very, very badly. Send the following people out for a Starbucks run:

- **Your sales agent:** The main function of a salesperson is to sell, not to troubleshoot issues. Sales staff may have a general idea of how telecom works, but technical details are frequently not their strong suit. Having a salesperson on the line during a circuit installation generally causes everything to take longer and runs the process off on tangents. You should greatly discourage the person who sold you the service from being on the call for the installation. Your sales rep has better things to do, and everything will move much smoother without his or her presence.

- **Any nontechnical people with strong opinions:** This category of people includes managers, directors, vice-presidents, and CEOs of your company. The installation of a circuit is an event that requires a great deal of diplomacy on all sides. If someone on the call begins attacking one of the technicians on the line because of a problem, the technician may quickly (and conveniently) prove the issue is not within the carrier's network and cease to troubleshoot the problem at that time. Only technical people with a calm demeanor should be allowed on an activation conference call. The emotional and uninformed will make the process take twice as long and not engender any good will with your carrier. And we can all use more good will.

Preparing for the Installation

The urgency of an installation dictates how much effort you put into preparing for it. In the perfect world, you prepare for every contingency as if the life of your company were riding on it. You probably don't have a ton of time to devote to planning. If you're replacing the dedicated circuits that are currently connected into your phone system with dedicated circuits from a new carrier, you must coordinate the *hot cut* thoroughly. If you are activating an additional circuit in a parallel cut, you don't have as much to fear.

From the moment you unplug your old carrier's circuits from your phone system until your new carrier activates your phone lines and you have clean calls passing back and forth on the lines, you have no phone service. I recommend that you do hot cuts like this after-hours to avoid negatively impacting your business. To prevent the hot cut from turning into a fiasco, you should work through the following checklist at least 24 hours before your scheduled installation:

1. **Confirm your installation time and date with your carrier.**

   Many carriers require you to schedule installation 24 to 48 hours in advance, just like you would schedule an appointment with your doctor. You can request a specific time, but if it is not available you may be given the next available time slot. This is an important detail, because you should schedule your hardware vendor based on the time you receive, not based on the time you originally had in mind.

   You carrier may be in a different time zone than you are, so you will need to confirm the time zone listed for your installation. If you are on the West coast and the installation department for your carrier is on the East coast, you could easily show up for your installation three hours late and be forced to reschedule. To make matters worse, some carriers send a confirmation e-mail with the date and time, but list the installation time based on Greenwich mean time (GMT), not a domestic time zone. You can use the Time and Date Web site to help if you need assistance (`www.timeanddate.com/worldclock/converter.html`).

2. **Confirm your installation time with your hardware vendor.**

   Hardware vendors have to conduct telecom triage just like everyone else in the industry. If the vendor is dealing with a huge customer whose system is completely down, its technicians may be occupied for a day or two bringing the service back to life. It's always good to confirm that the appointment is still on a day in advance, as well as confirm that you have the correct cellphone number and any other contact information you may need.

3. **Confirm that the local loop has been dropped, tested, and that you can find it.**

   Just because your local carrier dropped a new T-1 jack on the wall doesn't mean that the circuit is ready to install. The wiring from your building through the local network, and on to your long-distance carrier has to be completed and tested before you can schedule your activation. You should also confirm that the T-1 jacks are on the wall where you need them and that the circuit IDs for each jack match the confirmation documents you received from your long-distance carrier.

   If your long-distance carrier claims that the circuits have been dropped to your office, but you can't find them, jacks may have been placed in the MPOE *(main point of entry)* for your building. Your hardware vendor will need to extend the circuits into your phone room, so you will have to ask your local carrier to come and extend that wiring to your phone room. (For more about MPOEs and carrier responsibilities, please check out Chapter 4.)

4. **Confirm the order number and specifics with your carrier.**

   The order you initially sent to your carrier may have little resemblance to what was eventually delivered to your office. It is not unheard of to have the quantity of circuits, as well as their configuration, modified before the circuits are installed. You may have determined that you needed one more circuit or a different protocol would serve you better between the time the initial order was sent and the circuits were slated for installation. Quickly confirm the quantity of circuits that are to be turned up, as well as their general configuration.

5. **Confirm that your toll-free numbers are ready.**

   If you have any toll-free numbers that are being activated on the new circuits, you will need to confirm:

   - The RespOrgs have been released to your new carrier for existing numbers.
   - All new numbers you have are still on reserve.
   - The quantity of toll-free numbers to be activated to your circuit is the quantity you asked for.
   - The DNIS, ANI, and ANI Infodigits are ordered properly.

6. **Confirm that a conference bridge has been set up.**

   Don't take it for granted that someone is going to set up a *conference bridge.* A conference bridge is simply a setup that enables everyone to participate in a conference call. Bridges usually require a dial-in number and password. For simple installations that only include your hardware vendor, your carrier, and you, you can probably bring everyone together by using your phone system's conference-calling features. If you need another person on the line, your carrier or hardware vendor may be able to conference on another person. If you have more than three people required on an installation, it is best to have your carrier set up a conference call so everyone can dial in. In any event, everyone should know who is calling whom.

Writing an Installation Journal

A lot of things happen at the same time when a dedicated circuit is being set up. Technicians conduct tests, and nitty-gritty specifics of the circuit are discussed. I recommend keeping a journal of everything that takes place during the installation of your circuits. You may need to refer to it at some future date to clarify what really happened and refresh the memory of a forgetful, finger-pointing carrier or hardware vendor. Be sure to include the following:

- Date and time of the installation.
- The name, contact information, and title of everyone involved in the call, along with the contact's company affiliation.
- A running chronology of events. Include what each individual said or did. For example:
  - When each specific circuit was fully activated and ready to send and receive calls. You know your circuits are at that state if your carrier's installation technician tells you he has effectively *idled up* the circuits.
  - When each test call was attempted.
  - The results of each test call.
  - Specific troubleshooting tests that were attempted.
  - The results of the troubleshooting tests.
  - The moment each circuit was accepted.
  - The moment the installation was abandoned and rescheduled (and of course the reason for the abandonment).

Installing the Circuit

There is a standard procedure for activating a circuit. Every carrier has its own twist on how they do each step, but essentially, every dedicated circuit installation follows a normal progression.

Before everyone arrives on the conference call to activate your circuit, you are probably not connected to the new phone jacks, and your carrier will have the T-1 level of your circuit *busied out* (that means that if you connect to the jacks and try to call out over the circuit, you can't complete a phone call, or even receive dial tone). The technical term for the disposition of a dedicated circuit prior to installation is IMB, or *installation made busy* state — see the accompanying sidebar for more information about IMBs. So, you can't make any test calls until your carrier manually activates both the T-1s and the individual channels, called *DS-0s.*

Step 1: Introducing and reviewing

After everyone is on the conference bridge, the first thing you need to do is some general accounting and review. Some of the people on the call may be calling in on their cellphones and working in a wireless-unfriendly environment (most phone closets or server rooms are wireless vacuums), so you will need to write down everyone's name, company name, and phone number so you can call them back in case they get disconnected.

If I haven't already implied this throughout the chapter, let me be very clear: you are running this conference call; you need to take the lead by running through the details and getting the information you need. You're essentially coordinating the situation, so with your leadership (and barring any major technological glitches that are outside your control) the situation is likely to go smoothly. Or you could let the installation become a free-for-all or a circus by not being in charge. Your choice.

When everyone is accounted for, review the specifics of the order you are installing. This process should take no more than 30 seconds; run down

- The number of circuits you're about to install
- Their line coding and framing
- Their outpulse signal and start
- Their trunk group configuration
- The quantity of toll-free numbers
- The DNIS configuration

Welcome to your own private Bermuda Triangle: IMB

Prior to installation, every dedicated circuit will be in an *installation made busy* or *IMB* state. Busying out the circuits in the IMB state prevents the circuits from accessing your carrier's network and causing it to show alarms on your circuits. All carriers monitor circuits, and if your hardware isn't installed, your carrier's monitors see the circuit open and alarms are activated in your carrier's network. By placing your circuits in an IMB state, the network blocks your access, your circuits aren't seen as open, thereby stopping those alarms from blaring.

Whenever your carrier sees that your hardware is disconnected from the dedicated circuits on its network, an alarm sounds. Carriers hate alarms. The official, consistent response to an alarm is to place the circuit in an IMB state. That means that if you disconnect your circuits on Friday night to move some hardware from one phone room to another, and you plug them back in on Monday morning, you can expect to be blocked from your carrier's network. You will have to arrive on Monday and open up a trouble ticket with the carrier to have the dedicated circuits taken out of IMB. It doesn't matter if your circuit has been up for six days or six years; your carrier will eventually block your circuits and put them in IMB if you have them unplugged for any significant amount of time. Of course, the definition of significant varies from carrier to carrier and day to day. Sometimes you can go 12 hours before the IMB state is instituted. Other times you're sent into limbo after just two hours. You can call your carrier's customer service department on Friday and open a tech assist trouble ticket for Monday morning so that a technician is available to reactivate or correct any circuits that don't want to come back to life. Your circuit stills end up in IMB, but the trouble ticket provides someone to reactivate the circuit before you need it. I cover setting up *tech assist tickets* in greater detail in Chapter 13.

You can rattle this information off just like a laundry list (check out Chapter 8 if you have questions about line coding, framing, and outpulse signal and start peruse Chapter 9 for information about the RespOrg LOA, which includes trunk group information and DNIS), but the idea is that you give everyone a chance to validate what you are activating. If your carrier has only two circuits listed for installation (and you've got eight), or a different outpulse signal and start, it is better to find it now than after an hour and a half of testing.

Step 2: Connecting your hardware to your carrier

When everyone agrees on what's being installed, and nobody is confused about the order, connect your phone system to the new T-1 jack on the wall, listed with the circuit ID for your new long-distance carrier. After you plug the cable into the jack, connecting your phone system to the local loop, your carrier should activate the circuit by bringing it out of IMB. Your circuit is now live at the T-1 level, a fact that your carrier will confirm if your hardware appears on the network radar.

Your carrier will always turn up the highest level circuit first, and work his way down to the T-1 and the DS-0 level. If you ordered an OC-3, he will activate and test the OC-3 level first, and then the DS-3 level, the T-1 level, and finally the individual channel or DS-0 level. You can't activate the lower-level circuits until the upper-level circuits are running clean. The DS-0s will not come up if you are having problems at the T-1 level.

Step 3: Activating the individual channels

The individual DS-0s will be *activated* after the T-1 level is working fine. At this time, you know exactly how your installation is progressing. If the rep from your carrier tells you he sees all of your channels *up and idle*, you should be looking at a simple installation and you will be done soon. If your carrier is having problems either at the T-1 or DS-0 level, it may take a few hours to work through all the issues.

If problems arise, don't forget to write down what the problems are, what tests are run to resolve them, and so on.

Step 4: Making test calls

After the DS-0s are activated and everyone sees them in an idle state, you can make some test calls. This is simply a fancy telecom way of saying make a phone call. It is easier for everyone if you can isolate the exact channel you are dialing out on for your test calls, because it allows the installation tech at your carrier to monitor that circuit and watch you dial out. If your phone system doesn't allow you to designate a specific DS-0 to send out your call, that isn't a problem because your carrier can find the record of the call after it is made.

As long as you can grab one of the individual DS-0s, dial out, and the call completes without any static, echo, or being sent to a recorded message telling you that your call can't be completed as dialed, you have a successful test call.

Test every T-1 circuit you have, just to ensure that all lines are working properly. When all of your test calls complete fine, you can accept the circuit from your carrier and the carrier closes the dedicated circuit orders.

Step 5: Testing the toll-free numbers

Testing the inbound calls on a dedicated circuit is always done last. It would be a catastrophe if you had moved over all of your toll-free numbers on to your dedicated circuit only to find out that the circuit was configured incorrectly by your local carrier and won't be fixed for 48 hours. Toll-free numbers are very delicate creatures, and after you move them someplace they aren't supposed to be, it may take quite a bit of effort to move them back as they were intended.

If you're migrating existing toll-free numbers to a new circuit with a new carrier

If you're activating existing toll-free numbers that were migrated from another carrier, you need to have your carrier build a test toll-free number before you do anything. Your carrier should have an inventory of toll-free numbers for this very purpose. Together, you and the installation technician need to do the following:

1. **Technician configures one of the carrier's test toll-free numbers with the same DNIS digits, ANI delivery, and ANI Infodigits requirements that you're using for your toll-free numbers.**
2. **Technician points this fake number to your new circuit.**
3. **You call the test toll-free number.**
4. **You and the technician confirm that the test number is hitting your phone system, that the DNIS is being accepted, and that the call is being sent to the correct extension within your office.**
5. **If a problem is discovered, the technician makes corrections and you go back to Step 1 of this list before the toll-free numbers are updated to the route path.**

   Potential problems include incorrect information being sent in the DNIS stream, or an incorrect routing (meaning that the call isn't hitting your circuit).

At the time of installation, your new carrier must be in RespOrg control of the toll-free numbers; however, the carrier won't be receiving the traffic on its network yet. Only after the calls on the test toll-free number successfully complete should the carrier update the SMS database and have the calls routed onto its network. If the installation technician routes the calls onto the carrier's network before the test calls are terminating properly, he may not be able to send the calls back to route over the previous carrier in the event of a problem, essentially endangering your ability to accept calls from your toll-free numbers. At that time you would either have to accept that your toll-free numbers are down until your carrier can repair the problem, or escalate with your carrier to find someone in either the RespOrg department or toll-free routing department that can send the traffic for your toll-free numbers back to the previous carrier. During this process, it's very important for you to listen to the sequence of events so that you don't inadvertently let something happen that could prevent you from using your phone lines. If something bad happens to your toll-free numbers at this point, go to Chapter 9 for assistance and options for resolution.

If you're activating new toll-free numbers to a new dedicated circuit

If you are activating brand-new toll-free numbers, you can use one of your numbers as your test number because if the numbers are down for another 24 or 48 hours, it shouldn't really matter (the loss of toll-free number access that heretofore hadn't existed is inconvenient but doesn't threaten your business).

Your carrier's installation technician and your hardware vendor should be able to see the exact DNIS stream being sent back and forth. As long as the call is hitting your circuit, the technician should be able to correct the rest of the *handshaking issues* (issues that affect how your phone hardware and you carrier's hardware play well together).

Identifying Installation Problems

There are three general problems you may experience during an installation. The first problem stems from a lack of connectivity, the second stems from failing hardware, and the third stems from an issue with configuration.

Issues of connectivity

Connectivity issues are usually caused by problems with the physical wiring. If the wiring doesn't run without any gaps from your long-distance carrier (on a long-distance circuit), through your local carrier, and end up in your building, you will have a problem. There may be a dozen points between your carrier and your phone system where the wiring is cross-connected onto new circuits or through new hardware. If any of those connections were not completed, you will have an *opening* on your circuit. It is as if you have a broken water line. You may have good pipes at your home, and they have good pipes leaving the water company, but somewhere in the middle, things aren't connected.

Connectivity issues affect the entire circuit. If your T-1 is not cross-connected somewhere in the system, your carrier will not be able to detect your hardware, and you will not be able to see them either.

Issues of hardware failure

If a piece of hardware is failing somewhere in the system, you will be able to see your carrier, and they will be able to see you, but there will be errors on the circuit. These errors may result in you hearing static on all of your calls, the calls may disconnect prematurely, or you may not be able to complete a call. Fortunately, the same troubleshooting for continuity also works to isolate issues where someone's hardware is generating errors.

Configuration issues

Configuration issues are simple for a competent vendor to detect. As long as you are configured the same as your carrier, everything should work fine. The problem comes in when you have a vendor that does not know your hardware. If you ordered your circuit as B8ZS/ESF Loopstart, and your hardware vendor cannot confirm that your Newbridge 3624 Mainstreet (or whatever hardware you have) is configured that way, you need a new hardware vendor. If your technician is nice, he or she will slowly change all the settings to see what works for your channel bank, but that may take hours to reconfigure for each protocol. To avoid configuration issues, ensure that your hardware vendor can positively confirm all aspects of your hardware and change them if necessary.

Troubleshooting Continuity Issues

Before we begin troubleshooting, we should look at a standard hardware design for a dedicated circuit and see who is responsible for what section of the circuit.

Figure 10-1 shows the standard layout for a dedicated long-distance circuit, regardless of who ordered the local loop. The only variations you may have from this would be the location of your CSU (it may be inside your multiplexer). Also, the responsibility of the circuit from your local carrier may include an extended *demarc point* or jack in your phone room if the local carrier provided the inside wiring and not just the *network interface unit* (NIU).

You're responsible for all issues within your phone system through all the hardware and wiring to the NIU dropped by your local carrier. The local carrier is responsible from the NIU to the location where the local carrier hands off the circuit to your long-distance carrier at the Carrier Facilities Assignment (CFA) point, and your long-distance carrier is responsible for everything beyond that point. If you have any questions about the NIU, CFA, or POP listed in the diagram, check out your dedicated circuit options in Chapter 8 and look for hardware details (such as info on CSUs and multiplexers) in Chapter 4.

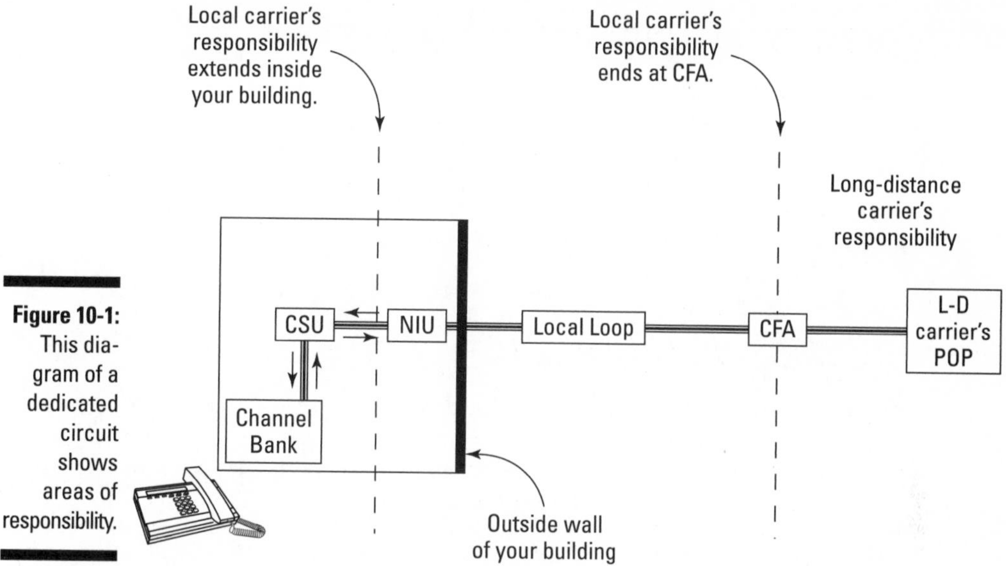

Figure 10-1: This diagram of a dedicated circuit shows areas of responsibility.

Any continuity issue or problem where you are taking errors on a circuit must be isolated down to the piece of hardware or carrier responsible for the issue. The standard procedure is to have your carrier test to the farthest point on the circuit, and then slowly step back until a point is reached where there is continuity without errors.

Asking for installation trouble resolution timelines

Trouble is handled with a bit more urgency during an installation. It may be handled more efficiently because the installation techs have more time to focus on pushing the repair than the customer care techs (who are handling multiple trouble tickets and are overloaded with escalations) usually have. If there is an issue with the local loop, you can expect some action to be taken in about three hours. If it is something within your carrier's network, they may be able to resolve it in 60 to 90 minutes.

Always ask for an approximate repair time frame for any issue you have during a dedicated circuit installation, because you may have disconnected your old carrier from your hardware and have no phone service until the issue is fixed. If the problem is going to take more than a few hours to resolve, you will have to decide whether you need to take your phone service back to your old carrier in the meantime. This is the only difference when troubleshooting a new circuit versus troubleshooting an active circuit. Skip to Chapter 13 for basics on troubleshooting dedicated outbound call issues.

Part IV
Taking Care of Your Telecom System

"It's the break we've been waiting for, Lieutenant. The thieves figured out how to turn on one of the stolen video conferencing monitors."

In this part . . .

Telecom is a club fight, and this maxim is most evident in troubleshooting. Carriers and vendors commonly become agitated and pass the buck. Unless you know more than the people you're talking to, you may end up accepting what you're told and run off to troubleshoot your issue down a blind alley.

By getting a methodical system for troubleshooting any phone service you may have, you can prevent yourself from being misled. Take the keys to the telecom kingdom! Part IV takes the essential pieces of all previous chapters and reassembles them so that you can resolve outbound and toll-free trouble issues. Starting with the basics of troubleshooting and the isolation of the variables in any phone call, you break down outbound and toll-free calls from a standard phone line. The final chapter covers the step-by-step troubleshooting of dedicated circuits, and toll-free calls that terminate on your dedicated circuits. Enjoy your freedom!

Chapter 11

Maintaining Your Telecom Services

In This Chapter

- Using general etiquette
- Reviewing types of calls
- Focusing on carrier-level problems
- Understanding the many types of call treatments
- Managing and monitoring your trouble tickets
- Troubleshooting international calls and faxes

The one thing I can guarantee about your phone service is that there will be a day when it won't work. The problem might be minor and short lived, or all encompassing and last for days. If the problem doesn't affect everyone in your area, the speed at which it's resolved is based on how quickly you provide accurate information to your carrier. The carrier might be unaware of any problem in your area until you report it.

This chapter covers the structure of the trouble-reporting network within your carrier, to help you understand what information the carrier needs. I also walk you through the basic information that telecom companies use to isolate problems, including a general overview of all the responsible parties handling a call. I explain how to determine whether you should report a problem as a dedicated or switched issue. Finally, I cover some tips on managing and tracking your issues.

Using good troubleshooting etiquette

If you aren't used to dealing with phone problems, it can be a very stressful and frustrating process. These conditions sometimes make people aggressive and emotional, two traits that don't help you as you resolve these issues with your carrier. As you go through the troubleshooting process, try to build goodwill with the carrier's representatives. They might run you through an automated system that takes two hours before you get to speak to a technician, but if the technician likes you when you finally do get to chat, he or she might let you call directly the next time. On the other hand, if you call the carrier with an attitude and venom in every word, the representative might be less than compassionate or concerned about the outcome of your problem.

Understanding Troubleshooting Basics

When you begin troubleshooting a problem, you are under the assumption that something in your phone system isn't performing at an acceptable level. For example, you might have echo when you call your grandmother or static when you call work. You might not be able to get through to your girlfriend because someone in the world of telecom thinks that her number has been disconnected or is no longer in service. The issue can affect all of your calls, both inbound and outbound, or just outbound calls to a specific city or state. Before you can begin testing a specific issue, you have to narrow down the problem.

When troubleshooting your issues, it's not your responsibility to correct the problem. Your main responsibility is to accurately report the issue. You can speed the process along by doing some troubleshooting yourself, but generally only your carriers or your hardware vendor can fix a problem. By isolating the issue to a single source, either a carrier or a piece of hardware, you can reduce the troubleshooting time by hours or possibly days.

The first step in troubleshooting is to systematically isolate the potential causes and eliminate them one by one. You are aided in this process by simply observing whether the issue affects a variety of calls or just a small group of them. Every type of call, whether it's inbound to your toll-free number or outbound to a local business, is handled by a different group of carriers. Any issue that isn't intermittent is easy to track down and troubleshoot by checking a handful of call types. There are only six varieties of call, so by figuring out the characteristics of the issue, you can either eliminate or indicate the source of the problem.

Switched outbound long-distance call

A switched outbound long-distance call is one of the most basic calls that people make every day. You make this call from a regular phone line, just like the one at your house, to a phone number that is outside of your LATA (Local Access Transport Area), state, or country. The distilled version of the call is represented in Figure 11-1.

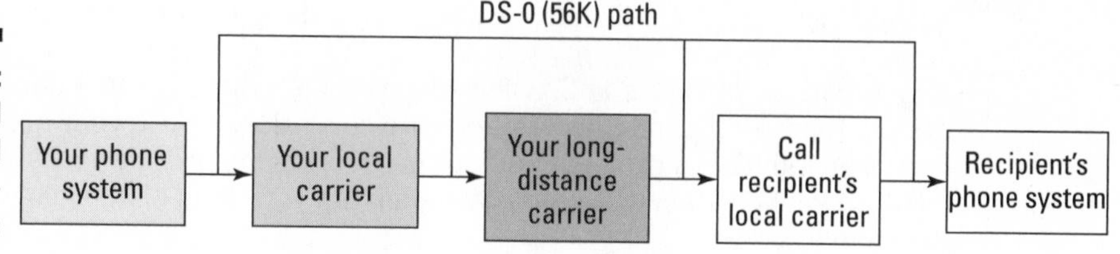

Figure 11-1: A switched outbound long-distance call.

You can see in Figure 11-1 how the call starts with your phone system (your phone system may be as simple as a single phone attached to a jack), and then progresses from your local carrier, to your long-distance carrier, to the local carrier that handles the service for the person you are calling, and finally to the phone you are actually calling. This is one of the most basic calls that you make, and probably the type of call that experiences the greatest quantity of issues.

Before you can pinpoint a problem to any variable in the diagram, you need to compare the call to at least one of the remaining five call types.

The *public switched telephone network*, or *PSTN*, is a generic term for all the long-distance carrier (Points of Presence) POPs, and hardware that routes and directs phone calls, as well as the local carrier central offices (COs), and all of their hardware that routes and directs phone calls. While a call is being sent through the PSTN, it is aggregated onto circuits with other calls for part of the path it takes from origin to termination. Imagine the PSTN is like a huge freeway for phone calls. All calls use the same roads, sometimes only until the nearest offramp, and sometimes for hundreds of miles. If a problem on a call happens while it's being routed through the PSTN, the issue is said to be in the *switched network*, even if the call originally started, or ended, on a dedicated circuit.

Switched inbound long-distance call

A switched inbound long-distance call is a very helpful call type because it eliminates your long-distance carrier as a potential cause of any issue. The long-distance portion of the PSTN is filled with hardware that routes and maintains the quality of your call. The greater the distance the call must travel through the PSTN, the more hardware is required to maintain the call. More hardware, of course, means that there are more potential points of failure, and so you are more likely to encounter a problem in the 28 or 2,800 miles where your long-distance carrier is handling your call.

As you can see in Figure 11-2, when someone calls directly into your phone system (not using a toll-free number), the caller doesn't use your long-distance carrier. If you have a problem on both outbound long-distance calls and inbound long-distance calls, you can eliminate your long-distance carrier as the source.

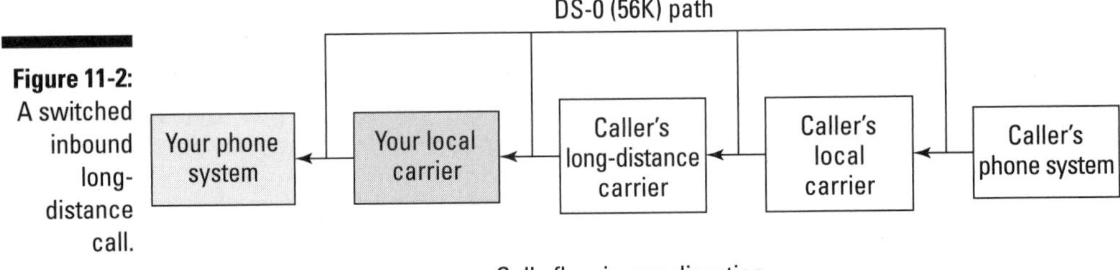

Figure 11-2: A switched inbound long-distance call.

Switched local call (inbound or outbound)

A switched local call uses the same carrier and interacts with the same hardware (carrier hardware and your phone system) for both outbound and inbound calls. It doesn't matter if you call your friend across the street or if someone calls you; the only carrier that generally ever sees your call is the local carrier you both use. Figure 11-3 offers a visual representation of this call type.

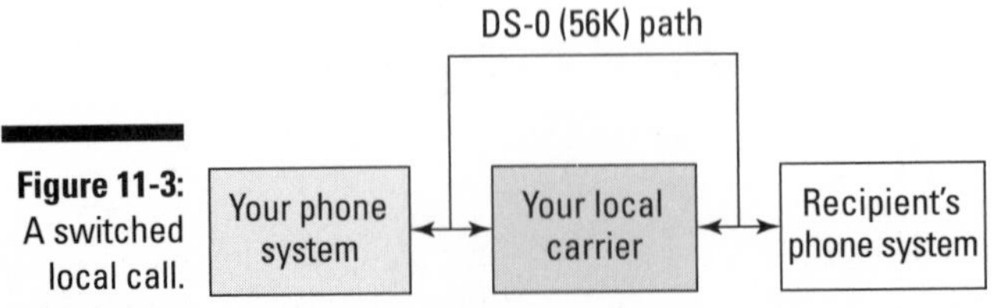

Figure 11-3: A switched local call.

You can see in Figure 11-3 that the call barely enters the PSTN and might pass through only one central office. If you have a problem with a local call, the possible sources are your phone system and the local carrier.

Dialing someone else's toll-free number

When you dial someone else's toll-free number, the only piece of the call that you have any responsibility for is your phone system. Your local carrier is responsible for identifying the carrier that receives the traffic, and then your local carrier forwards the call to that network. Figure 11-4 shows the path the call takes when you dial someone else's toll-free number.

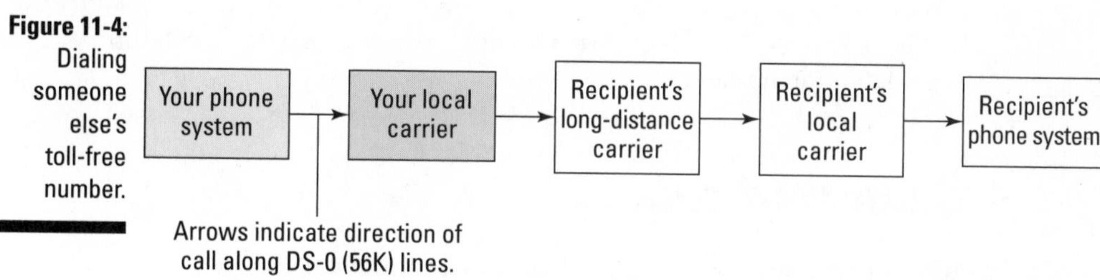

Figure 11-4: Dialing someone else's toll-free number.

If you have a problem calling someone else's toll-free number, but all of your other calls work fine, the problem probably has something to do with either the carrier that handles the traffic of the toll-free number, or the local carrier that terminates the call. Identifying which carrier is the problem is covered in the step-by-step troubleshooting section for toll-free numbers in Chapter 12.

Many companies use the same long-distance carriers. If you use Sprint for your long distance and you dial the toll-free number of a company that also uses Sprint for long distance, you can possibly see a similar problem on your outbound calls. In this case, you should focus on troubleshooting your outbound long-distance calls on the Sprint network. When Sprint resolves your outbound issue, the toll-free problem for the company you are dialing to will probably also be fixed.

Someone is dialing your switched toll-free number

The last switched call type is an inbound call to your toll-free number. The diagram for this call, shown in Figure 11-5, looks quite a bit like Figure 11-1. It includes the same path.

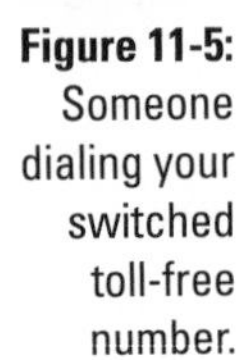

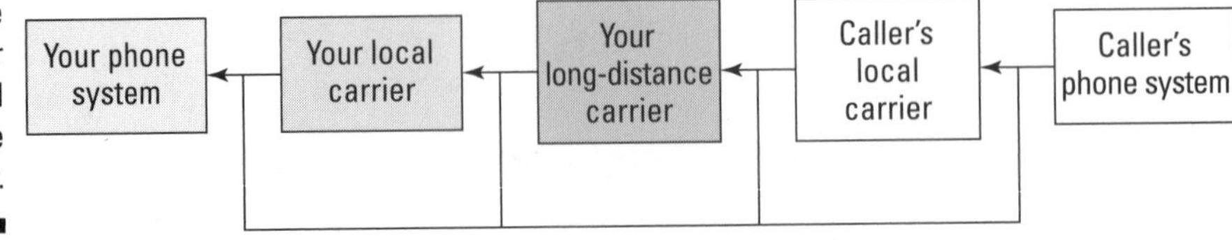

Figure 11-5: Someone dialing your switched toll-free number.

Arrows demonstrate path of call along DS-0(56K) lines.

The main aspect that separates a toll-free issue from a regular outbound issue is the fact that the call is being sent to your long-distance carrier based on a database check done by the originating local carrier. Aside from how the call is sent to your long-distance carrier, the path the call takes encountersall the same carriers and hardware it would if you had made an outbound call to the origination number.

Dedicated calling

You can compare calls over your dedicated circuits with switched calls in order to eliminate your local carrier as a possible source of the problem. Figure 11-6 shows how the local loop of your dedicated circuit passes through your local carrier, but doesn't really interact with it.

Keeping the switched network in mind when you have a problem on a dedicated circuit

When you report trouble on your dedicated circuit, keep in mind that the local loop is the dedicated section your calls pass through. If you have constant echo, static, dropped calls, or if the entire circuit is completely down, you should refer to your circuit when you open the trouble ticket. If you have any problems that you can duplicate by dialing out on a regular phone line over your carrier's switched network, such as calls failing to a certain area code, fax completion problems, static, or echo, you should open a trouble ticket with your carrier based on the call from your regular phone line. If the problem exists on both your switched and dedicated calls, but you open the ticket as a dedicated issue, the problem will be harder to diagnose.

Carriers *do* assign greater priority to dedicated circuits, but they also push to intrusively test your circuit as the first step in troubleshooting. The intrusive testing process prevents you from using any channel on your circuit during the testing. If you can open the trouble ticket on a call from a regular phone line, your carrier won't be distracted by your T-1 and waste time testing your local loop.

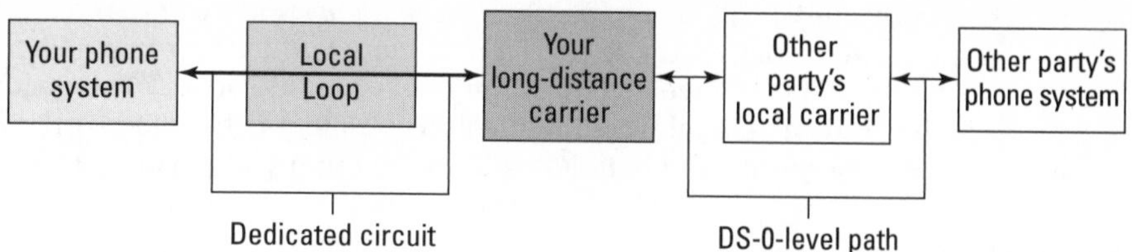

Figure 11-6: Dedicated calling process.

There are only three elements that interact at the individual channel (DS-0) level on a dedicated circuit:

- **Your channel bank or multiplexer:** This is the piece of hardware at your office that breaks the dedicated T-1 circuit down into 24 useable DS-0 circuits that transmit and receive your phone calls.
- **Your carrier's switch:** The one piece of hardware in your carrier's switch that you should be concerned with is the section that functions like your multiplexer.
- **Echo cancellers:** These pieces of hardware eliminate the echo on your phone calls. They may be in your long-distance carrier's network between the POP and the local carrier, or within the local carrier network.

The next important area to note in the diagram is the point where the call enters your carrier's network to the point where it's delivered to the recipient's phone. Figure 11-7 shows an expanded view of a dedicated call.

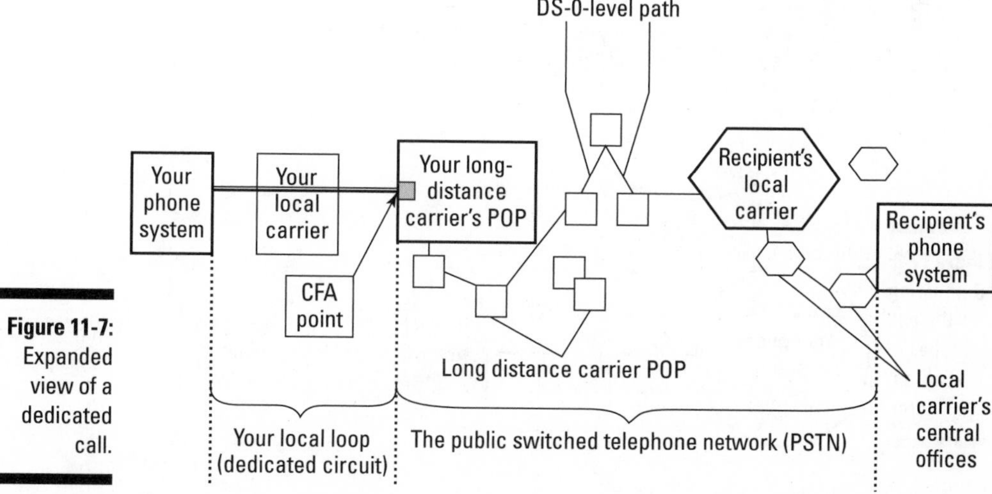

Figure 11-7: Expanded view of a dedicated call.

In Figure 11-7, you can now see that the local loop begins at your building and ends at the CFA *(Carrier Facilities Assignment)* point where the local loop enters your long-distance carrier's network within their POP. (Check out Chapter 8 if you have any question about the CFA point.) The majority of a dedicated call actually isn't dedicated at all, but is handled by the PSTN, where it's routed and handled with every other call in your carrier's network.

Comparing call types

Call type diagrams are great tools when you are troubleshooting. You should record the types of calls you have made and note whether they experienced the problem. Then you can refer to the diagrams and begin eliminating variables. For example, if you have static on local calls and inbound calls, but not on dedicated calls, you can distill the variables in the following manner, as shown in Figure 11-8.

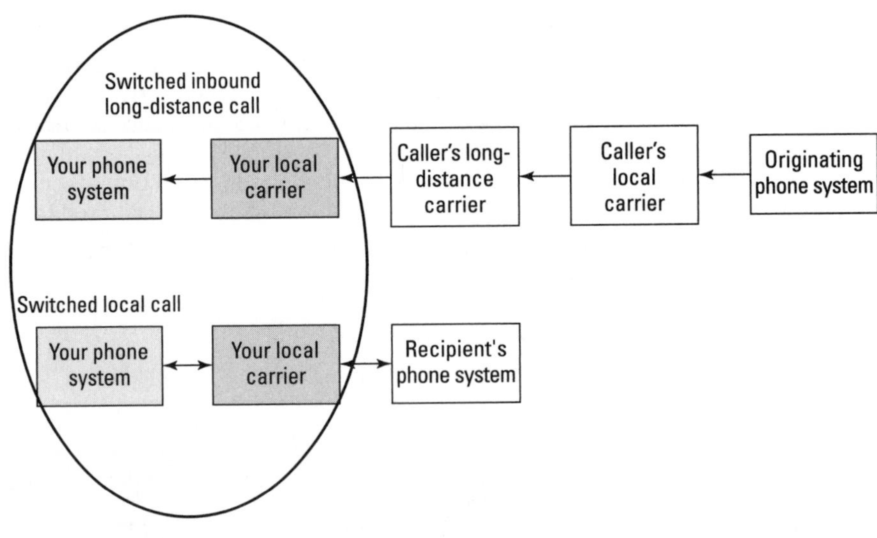

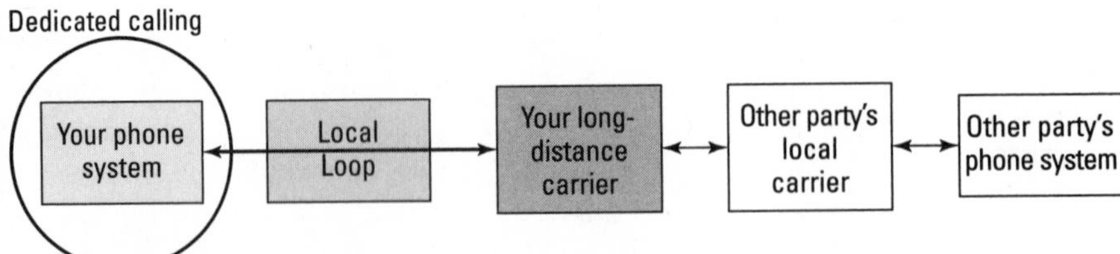

Figure 11-8: Comparing the call types in diagram form.

Keeping an open mind to the information

The speed at which a problem is resolved is directly related to how well you accept the results of your testing. It's not uncommon for people to place the calls side by side and have a good indication of where a problem is, but not troubleshoot that section first. This is very common when all the signs indicate that the issue is with your hardware. Open a *technician assistance ticket* (also called a *tech assist ticket*) with your carrier to schedule a time with their technician to walk you through testing and work with your hardware vendor. The help of your carrier's technician aids your hardware vendor in isolating issues within your phone system, but your first call should be to your hardware vendor if you suspect that is the site of your issue.

When troubleshooting, look for similar paths. For example, when you compare the diagrams of the inbound calls and the local calls, you see that there are only two variables in common: your phone system and the network of your local carrier. Local calls are commonly handled by your phone system, your local carrier, and the phone system of whomever you are calling. If you simply call another person in your local area and have the same problem, you can eliminate the hardware of the first person you called as the potential problem. The only two variables left for you to investigate are your phone system and your local carrier.

Every inbound long-distance call that comes from various companies in different sections of the world has its own local and long-distance carrier. Therefore, you can quickly eliminate the other long-distance carrier and the other phone system. The only variables that can possibly be causing a problem are either your phone system or your local carrier.

After you consider the call from your dedicated circuit that did not have the problem, you can eliminate your phone system as a possible source of the issue. As long as you are dialing out from the same phone system over both your dedicated circuit and your switched lines, it's very unlikely that your phone system is the source of the problem. You have now reduced the possible sources of the problem to just one entity (your local carrier). The next step is to call the local carrier and open a trouble ticket. It's helpful to let the local carrier know about the tests you have done and that the problem doesn't affect calls on your dedicated circuit, where they don't interact at DS-0 level.

Narrowing down carrier-level problems

If your problem persists on all your calls, but you've narrowed down the source to one of your carriers (eliminating hardware as a source), you can easily locate the issue and make sure that it gets repaired. If the problem doesn't affect all of your calls, you might need to focus more on the pattern of failures within the carrier.

There are three additional bits of information that can help save time when you open your trouble ticket. Ask the following questions:

- **Is the issue specific to a time of day?** Your carrier might be sending your calls over an overflow route that isn't very stable at peak network times between 10 a.m. and 3 p.m. If this is the case, the issue might be with a route that isn't used or monitored very often. If the problem happens intermittently during your peak times and not the peak times of your carrier, you might have a bottleneck in your own phone system. Time-of-day issues typically indicate a problem based on the volume of calls passing over a network, and identifying a problem this way can steer the technicians to a faster resolution. Share the information; it's good for everyone.
- **Is the issue geographically specific?** If you can call everywhere in America except the West coast, the state of South Carolina, LATA 730, the 305 area code, or any geographic region that you can identify in a telecom-based group, note this information to your carrier. If there has been a large outage affecting all calls to the West coast, your carrier might not know this until you call.
- **Is the issue intermittent?** If the problem occurs randomly, regardless of time of day or geography, get ready to settle in for a long troubleshooting process. The first thing to do about intermittent issues is determine the percentage of calls affected. Issues that affect about 50 percent of your calls are not that difficult to find, because if your carrier watches 10 test calls, the problem is bound to show up in around half of them, making it easier to research and repair. If the problem affects 5 or 10 percent of your calls, you may make 10 or 20 test calls before one call is affected. When less than 10 percent of your calls are affected by a problem, you enter the world of the needle in the haystack. As long as both you and your carrier stay focused and have the time to devote to troubleshooting, you will eventually find the problem. If the issue is low on your priority list, it might persist for weeks or months.

Getting the Most from Your Carrier's Troubleshooting Department

Most carriers have a two-tier structure for handling problems. The first tier of people you meet are the entry-level customer service folks. These people generally work from a script that asks you specific questions to qualify your issue. After a customer service agent has all the information he or she needs, you're given a trouble ticket number for tracking purposes.

If the first tier of customer service can't resolve the problem, the trouble ticket is sent to the network technicians, who make up the next level of support. These people can manipulate the network, update switches, perform tests, and are empowered to fix the complex things that go wrong. These are the people you want to speak with when you have a complex issue.

If you have a difficult and intricate issue, give the first-level customer service people just enough information to open the ticket and ask to chat with a technician. If you begin trying to tell the customer service rep all the minutiae of the problem, it might get lost in translation from you, to them, to the notes, to the tech. It's always safer to simply get the ticket open, press for a tech, and then explain it all to the tech.

There are three points to keep in mind when chatting with the first-tier customer service staff:

- **The rep you talk to is probably working off a script that has required information. You can't proceed without going through this stuff.** If you don't have all the information you need, things won't go very far, so make sure you gather as many details as you can before you pick up the phone. For example, if you have a problem dialing a phone number in Alaska, but you don't know the number you dialed because you lost your notes, the carrier can't open the trouble ticket.
- **All the information you give guides how your problem is handled.** If you tell the customer care rep that your dedicated circuit can't dial to area code 414, the ticket will be sent to the dedicated department to test your T-1 lines. If the same problem happens when dialing phone numbers in the 414 area code from your switched phone lines, the trouble ticket is sent to the department in your carrier devoted to handling calls from the PSTN in the 414 area code to determine why that area code is failing.

  It is always preferable to open a trouble ticket as an outbound switched problem if it affects that call type. This option is always the most direct route to resolution and prevents your carrier from being distracted by inconsequential aspects of the problem. The same holds true for an issue that affects both your inbound and outbound toll-free calls to a specific area. Opening a trouble ticket as a toll-free problem focuses your carrier on the toll-free aspect of the call (SMS database construction, areas of coverage, and nuances that are particular to toll-free numbers), and not on the overall network (which is actually the source of the problem).

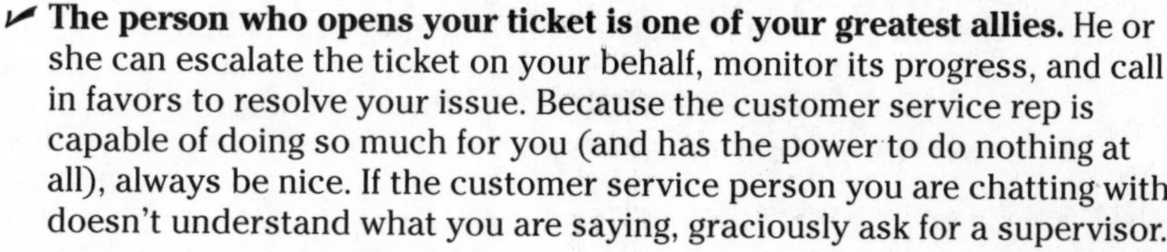

- **The person who opens your ticket is one of your greatest allies.** He or she can escalate the ticket on your behalf, monitor its progress, and call in favors to resolve your issue. Because the customer service rep is capable of doing so much for you (and has the power to do nothing at all), always be nice. If the customer service person you are chatting with doesn't understand what you are saying, graciously ask for a supervisor.

Understanding the translating and routing of calls

Every carrier receiving a call must determine where to send the call; then it must send it. These two activities represent two of the most common areas causing the failure of calls. *Translation* is the process a carrier undertakes to identify the destination of the network that must receive the call. Long-distance networks don't send calls all the way to the end telephone receiving the call; the local carrier handles that job. The long-distance carrier sends the call to the local carrier's specified *central office* (CO). If the local carrier decides to use a different CO to ring your phone, but your long-distance carrier is not aware of the change, your long-distance carrier sends the call to the wrong CO and the call fails. This problem is corrected by the translations department at your long-distance carrier, because the translations of the number you dialed to the correct CO must be resolved.

When a carrier's network determines where to terminate a call (by choosing the correct local carrier CO) the carrier has to deliver the call to that location. The *route* is the path the call takes from the moment it enters the carrier's network to the point it leaves the carrier's network. The routing department within each carrier monitors these routes and prevents them from being congested. If one main circuit fails, calls proceed through a secondary or tertiary route set up by the routing department. If a large outage occurs, or calls fail on their way to the correct local carrier CO, the routing department identifies the fault in the network and repairs it.

Identifying your call treatment

When you call to report a problem with your phone service, the rep at your carrier asks you to describe it. The more specifically you can describe the issue, the easier it is to find and repair. The rep might directly ask you for the *call treatment,* or the symptoms of the failed, or substandard call. This is where telecom can seem like a foreign language. You know what is happening, but putting it into words that can be understood might not be that simple. The main call treatments you will encounter are listed in the following sections.

Understanding why "your call cannot be completed as dialed"

If you hear a recording that tells you that your call can't be completed as dialed, you may have misdialed the number, the area code of the number you are dialing having changed, or a translation problem may have occurred in a carrier network somewhere along the line. This recording generally indicates that the problem affects an individual phone number. Your ability to dial other numbers in the state, town, or country is usually not impacted. Try the number again, check the digits and the area code, and then make the call to your carrier's customer service.

Understanding why "the number you have called has been disconnected or is no longer in service"

A recording that says that the number has been disconnected or is no longer in service might be legitimate; the number has simply been disconnected. If you know it's not disconnected, either you misdialed the number, or there is a translation issue somewhere.

Handling an "all circuits are busy" message

On rare occasions you may hear a recording that tells you that all circuits are busy. This recording rarely means that all the circuits available in your carrier's network are occupied, unless you are trying to call your Mom at 9 a.m. on Mother's Day, along with everyone else in the world. This recording is generally played when your carrier has an outage of some sort and a portion of their network is down. If a backhoe accidentally cuts through a phone line, thereby taking down phone service for the entire city, you will probably hear this recording when you try to dial out (the carrier unfortunately doesn't have a *sorry, but our main phone line has been cut by a backhoe* recording).

Listening for tones and tags

Tones and *tags* are supplemental sounds or recordings and are generally attached to a standard recording. Listen for *tri-tones* that are played before recordings by a local carrier. They are generally three ascending notes that sound like they come from a cheap synthesizer — very shrill and high pitched.

The tags are more important because they frequently list the switch that is playing the message. If you hear a recording of *your call cannot be completed as dialed — fifteen dash two,* you know that switch 15 on your carrier's network probably has a problem. If you hear tones and tags during a recording, be sure to note them, because alerting a technician to the specific tags can shorten the amount of time it takes to solve your problem. Instead of tracking down the failed call, the technicians can go directly to switch 15 and analyze it. The number after the dash may correspond to the recording played (for example, the two in this example may mean *cannot be completed as dialed*). On the other hand, the number may have no significance at all. Every carrier has its own system for the tags played. The tags may mean nothing to some carriers, but to others they can be significant. I say you should always have more information than less.

Understanding the fast busy signal

A fast busy signal is a busy signal that sounds twice as fast as the normal busy signal. You will probably hear a fast busy signal when part of your carrier's network is down (the pesky backhoe again — see "Handling an 'all circuits are busy' message," earlier in this chapter), so your call can't be completed.

Handling echo, echo, echo

Usually, only one person on a call hears an echo. Carriers have specific pieces of hardware installed throughout their network called *echo cancellers* (or *echo cans*) to eliminate echoes on calls. These devices can fail over time, be mis-optioned, or mistakenly installed backwards. The interesting thing is that if one person hears an echo on a call, it's probably the result of a bad echo can on the other side of the call. Sometimes, both people can hear the echo, but it's not as common.

Echo doesn't manifest itself in a way that's immediately visible to the technicians at your carrier. If your call fails to a fast busy signal, your carrier can pull the call record and find the switch that killed the call. Some issues, such as dropped calls or static, are visible in a circuit's performance report, which indicates that there has been an electrical or protocol-related anomaly. Echo on a call, however, doesn't leave a trail of breadcrumbs and is therefore a difficult issue to isolate and repair.

The reverberation of voice that people hear and classify as echo might also be the result of something other than a failing or mis-optioned echo can. This other type of echo is referred to as *hot levels*, and is the result of the over-amplification of a voice in the transmission of the call. If you didn't have an equalizer in your stereo and you cranked the volume up as far as it could go, your sound quality would be horrible and you would experience *reverberation* (or what you might call echo). To compensate for volume loss as calls are transmitted across hundreds of miles over circuits within the network, the switches in the POPs will boost the sound a fraction of a decibel, what is called *padding the call,* to keep the signal strong. The problem is that if the carrier amplifies the sound too much, you will hear echo. What is worse is that *hot levels* have cumulative effects.

If a call travels 15 miles and is padded 0.4 decibels in the two pieces of hardware it encounters, you won't hear echo. In fact, as long as the call has fewer than two decibels of padding from start to finish, you won't have a problem. If the call is traveling from coast to coast and each of the six long-distance POPs bumps the volume up 0.4 decibels, the call now has 3.2 additional decibels in padding. You can bet you will hear echo. The cumulative nature of hot levels makes it difficult to find and correct. In situations like this, the issue might persist for weeks before the carrier finds all the offending switches and reduces the padding.

Standing up to static

Only one side of a call usually hears static, but it does affect both incoming and outgoing calls. It's generally caused by a piece of hardware slowly dying in the network. The static can be very minor to begin with, but grows over time until you can't hear the person you're calling, or they can't hear you. As the issue evolves, the piece of hardware will eventually die and all of your calls will fail. Don't feel bad; the bigger the issue, the easier it is to find. When the static is noticeable on every call, it becomes simple to track it down and replace the affected hardware.

Echo might be invisible to the technicians monitoring a network, but generally static isn't. Almost all circuits in the U.S. are capable of being monitored for quality. This doesn't mean someone from your phone company is listening in on calls to ensure that they sound clean, but computer files can capture any unexpected electronic or protocol activity on a circuit. Static can be classified as *unexpected electric activity,* as it's some device experiencing an electrical short. The *performance monitors (PMs)* are diagnostic files for a circuit that record the electric and protocol anomalies. The challenge with static is finding the correct span causing the problem. If you're dealing with an intermittent issue affecting five percent of your calls somewhere in the switched network for your carrier, the possible locations of the failing hardware can seem endless.

Dealing with dead air

Dead air is a term that refers to the phenomenon when you hear nothing on the other end after you dial a phone number. You don't hear the dial tone anymore, but you also don't hear any ringing; you just hear nothing. When you hear dead air, stay on the call for 30 to 60 seconds; you'll probably hear a fast busy signal if you wait long enough. Dead air is generally caused by a translation or routing problem that caused your call to be transferred to a piece of hardware or a circuit that no longer exists. Because the hardware no longer exists, nothing is there to send you a polite recording, busy signal, or anything. All you get is dead air.

Dead air isn't the same thing as post dial delay (PDD). PDD is the silence you hear for a few seconds before you hear the ringing. Every call has some PDD, albeit it might last only one second. International calls are notorious for long PDD; anywhere from 15to 30 seconds may pass before you hear the phone ring on the other end.

Getting around clipping

Clipping is the technical opposite of echo. With echo, a voice is repeated; with clipping, sections of the voice transmission are lost. Instead of hearing the conversation in full, you hear only sections of each word, because a few milliseconds are lost every second. This problem is solved just like echo or hot levels, but instead of having too much volume on the call, you have insufficient volume. The troubleshooting process for clipping is the same as any quality issue and requires both patience and a large quantity of call examples.

Dancing around dropped calls

Phone calls that are disconnected before either person hangs up are deemed *dropped calls.* If your phone system loses power while you are talking, it will drop your call. The same thing happens if you are calling over a dedicated circuit that suddenly fails or takes an electrical hit. Dropped calls are researched by your carrier and the exact cause of the disconnection is identified by a *disconnect code* passed through the network when the call ends. The disconnect code identifies whether the call was dropped by the origination side of the call, the termination side of the call, or an unknown event in the carrier's network. Dropped calls are generally the result of a failing piece of hardware and typically become more frequent until the source of the problem is evident. If your carrier is causing the disconnections, its technicians will find the failing hardware by protocol failures in the protocol monitors of the circuit.

Handling aberrant recordings

Each carrier has a few standard recordings; you don't hear a ton of variety. Any other recordings you hear probably come from your phone system. Your carrier doesn't have any recordings on file that say, "We regret that you were unable to access an outside line," or "Your long-distance carrier is currently rejecting your call."

If you receive any recording that refers to your carrier, it probably was not made by the carrier's network. You might want to check with the carrier first to ensure the message isn't in its play list, but then you need to investigate, either with your phone system, or with the phone system of the number you dialed.

Working around incomplete dialing sequences

If you dial a long-distance phone number and the call is dropped after you press six or seven of the digits, you have two possible sources to check:

- **Incomplete dialing sequence caused by local carrier:** If your local carrier perceives that the number you are dialing is a local call, it might try to complete it after the seventh digit. If the call is actually a long-distance one, you need to take this issue up with your local carrier immediately.
- **Incomplete dialing sequence caused by your phone system:** This situation can also occur on some phone systems when international numbers are dialed. Your phone system might limit you to dialing ten digits. Because the international prefix of 011 already takes up three digits, and then possibly two to six digits for just the country and city code, you can easily reach the limit set by your phone system. Your call is dropped even before it leaves the office. You need to talk to your hardware vendor so that your phone configuration can be changed.

If you are dialing from a switched phone line, you must be able to complete dialing all the digits of a phone number before your local carrier processes your call and sends it to your long-distance carrier. If you have a dedicated circuit, check your phone system and then contact your long-distance carrier. The long-distance carrier can watch you dial the phone number and see every digit you enter. More important, the technician can determine whether the hang up on the call is being initiated by your phone system or by the network. When you know the cause of the disconnection, you know where to turn to resolve it.

Providing a call example

A *call example* contains detailed information that allows your carrier to follow the call's path from the moment you first dialed the number to the point the call failed. As technical as the idea sounds, a call example is just the information you write down about a failed call. After you dial out and get a *cannot be completed as dialed* recording, dial the number again and write down the necessary information (see the following sections for more information about what to include in your notes). When the carrier finds the call's end point, the technician can begin correcting the issue.

Call examples function to not only tell the technicians where to look for the problem, but also to allow the customer service rep to categorize the issue. Depending on the information you provide, the customer service rep will send your issue to a specific department for repair.

Call examples might not be easy to come by in some instances. If you are calling a number you dial often and the call fails, you will know all the information required to open a trouble ticket. The challenge comes in when customers dialing in on your toll-free number have an issue. Customers might not have your direct phone line to tell you that they couldn't get through and report the issue. Even if they do get through to you, it's not common to begin your conversation with a quiz about the specifics of a failed call attempt. As a result, you might have to ask one of your customers to make test calls for you. The specific information your carrier needs is listed in the following sections.

Call examples have a shelf life of about 24 hours. The specific information about how a call is routed is kept in your carrier's switches for a finite amount of time before it's overwritten with new, more recent calls. If an issue crops up on Friday at 5 p.m., you need your carrier immediately. If you try to provide the call example from Friday when you come into the office on Monday, your carrier will probably reject it and ask you to give them a newer one.

The date and time of call

Every call is logged into your carrier's switches by the origination time. If you made four calls to a phone number and only two of them failed, be sure to give your technician as much information as possible to help differentiate the completed calls from the failed ones. If you provide one call example and don't mention the other three attempts, the technician might find one of the completed calls and close the ticket because their research indicated that the call didn't fail.

There are four time zones in North America, plus Alaska and Hawaii, so be sure to identify the time zone when you provide a call example. If you don't tell your carrier that the call was made at 8 a.m. eastern standard time, the customer service person might record the time by using a local time zone.

The origination phone number

You might be one of 10,000 people who call a specific phone number every day. The only way to isolate your call from all the others is by referencing it back to your phone number.

If you are calling from a phone system that randomly assigns a line as you dial out, you might not know the originating phone number. This isn't a problem. Generally, the outbound phone lines you dial from are provided in sequence, so simply use any of the numbers as your origination, and then tell your technician that you are dialing from a phone system. The technician will note this fact in the ticket. When the call examples are found, the technician can trace the call back to your office by matching the area code and the first three digits of your phone number.

The number you dialed and the call treatment

If you made 5,000 calls in the past three hours, and one of them failed, your carrier needs to know which one out of the 5,000 numbers you dialed was the problem child. Seems reasonable, yes? The phone number you dialed gives your carrier an idea of the geographic area you are dialing into and is essential for tracking down the problem.

See "Identifying your call treatment," earlier in this chapter, for more on what information you should include when you describe the call treatment. Saying that the phone call "just didn't work" . . . just doesn't work.

Understanding when to provide multiple call examples

A phone call has many paths it can take to reach its destination. Depending on fluctuations in the capacity of the network between the two points, you could make 10 calls to the same number in 15 minutes and your calls may never take the exact same route twice. Understanding the complexities of the phone systems is the key to resolving intermittent issues. The larger a problem, the easier it is to track down and repair. If you hear dead air, static, or echo on 5 percent of all your calls, you need to provide as many call examples as possible.

Whenever you have an intermittent issue, it's very helpful to provide to your carrier clean call examples in addition to problem calls. This will allow the technician to review all the calls and begin comparing the individual circuits the calls took. After the technician eliminates all the circuits on the clean calls and isolates any remaining similar circuits on the affected calls, he or she will be more than halfway through troubleshooting your problem.

Managing Your Trouble Tickets

Many carriers are efficient, but the reality is that you are one of the tens, hundreds, thousands, or hundreds of thousands of customers the carrier has. It's common to have a 2,000-to-1 ratio of customers to customer service reps at a carrier. If you have dedicated circuits and bill several thousand dollars a month, you might receive some added support, but the ratio is rarely much better for big-time customers than it is for individuals or small businesses. With this information in mind, you can easily understand that your trouble tickets will get quicker resolution if you manage them yourself.

The format you use to manage your trouble tickets can be as complex and structured as you want. You can build a large relational database that tracks trouble tickets by carrier, issue type, start date, mean time to repair, and resolution, or you can go for something less structured. The least-technical way to manage your trouble tickets is to simply write down your call example and all the information you need to open the ticket on a piece of paper, and update it with times you called in and the status you were given. This information allows you to track how long the ticket has been open, and to track its progress. Of course, you might want to type this information and save it electronically just in case it gets thrown away with the other scraps on your desk.

If telecom isn't your main job, you might only open one trouble ticket a year, but if you expect to open one trouble ticket every month or week, consider creating your own trouble ticket form. In fact, I have included a really cool trouble ticket form in the Cheat Sheet of this book to get you started.

Understanding the timelines

Carriers have their own standard timelines for responding to issues, based on the severity of the issue. If you have one phone number in Minnesota that you can't get through to, the carrier's internal policy may be to provide you with a callback on the status of the issue within four hours. If the DS-3 circuit is down and you have no phone service, the standard time for a callback might be two hours.

Carriers typically won't give you an escalation until they have failed to respond to you in the standard interval, although local carriers are much stricter on these timelines than long-distance carriers.

That doesn't mean that your carrier will solve your problem in two to four hours; it simply means that you will receive a callback from a technician in that amount of time. I do recommend escalating trouble tickets if they don't experience any progress. If you have additional information about a problem, you can always call back into your carrier and update the ticket with the new call example or results of a test. This is a very good way to ask for updates without sounding pushy.

Coping with large outages

There isn't a network in North America that has not had a substantial outage. The source might be a *fiber cut,* a computer virus, or simply the growing pains of integrating a new technology like *VoIP* into their POPs. (By the way, a fiber cut is exactly what it sounds like — a cut cable causes a catastrophic loss of phone service.) When these problems hit, they are big, fast, and will generally take at least five hours to fix. Software issues that clog a carrier's internal network can take anywhere from one hour to five days to completely fix and are difficult to pin down. The mercurial nature of software issues makes any estimated time to repair dodgy, at best. Your best bet in these scenarios is simply to call in every hour or so for updates. The story can and will change as it goes along, and you will at least be informed about the progress.

A day in the life of a fiber cut

Fiber cuts are more tangible than other issues, so carriers can typically provide realistic timelines for repair. The greatest anxiety is waiting for the technicians to be dispatched and find the cut. If the cut occurred close to a large town, the technicians might be on site in 30 minutes. If it was in a rural area, it might take three hours. After technicians are on site, your carrier should be able to give you an estimate on the time to repair. For a simple fiber cut, the technicians have to trace the cables back from the split and dig down to find an undamaged section of the fiber to splice on a new section of cable. When one side is connected to the new cable, the technicians then have to dig another trench on the other side of the cut to expose the fiber and then begin splicing the new section of cable onto it. If you estimate one or two hours per side to dig each trench and splice the new cable, you should be in the right ballpark. Now, if there are extenuating circumstances and the fiber cut is the result of flooding that eroded a hillside, which exposed and shredded a mile-long section of fiber, you can expect a much greater time for repair.

Troubleshooting International Calls

From your perspective, troubleshooting an international call is just like troubleshooting a domestic call. You can run through the switched outbound troubleshooting steps in Chapter 12 and isolate the issue just as if you were calling Newark, New Jersey. The only twist to troubleshooting international calls is that you might find some interesting similarities when you try your call over other carriers.

The reason for call treatment similarities is because your long-distance carrier doesn't use its own network to complete calls to every country in the world. I can guarantee you that MCI, Sprint, and AT&T don't own all the cables, hardware, and have a staff of technicians around the world to connect your calls into Senegal, Papua New Guinea, and India. Your long-distance provider uses an *underlying carrier,* a company specifically designed for delivering international calls from the U.S. to a specific country or region in the world. The interesting thing is that there are only so many underlying carriers that provide service into each country, and every large domestic long-distance carrier probably has a contract for service with every large underlying carrier. In other words, more than one long-distance carrier is using the same path to complete calls to Gifu, Japan, or Prague, Czech Republic.

Carriers monitor their completion rates daily to every country in the world. If you try hard enough, your long-distance carrier might even send you the list of what they consider acceptable completion rates. Don't be shocked if you see that a completion rate of 60 percent to Western Europe is acceptable; the rate drops to around 7 percent or less for some African countries.

Your long-distance carrier can route your international calls over several underlying carriers. The choice of underlying carrier depends on the underlying carrier's completion ratios compared with all the other carrier choices at that time, as well as the price you are paying for your international calls. Some carriers have a premium group of underlying carriers available for international calls, but they can't place you on that group of carriers because they would lose money. If your business is focused on international calling, you might be better served by paying a few pennies more per minute for your calls, if you can realize a better call quality or completion rate. If you are opening more than one trouble ticket every few months on international issues, speak to your carrier about a better route.

Resolving International Fax Issues

International calls are prone to having quality and completion issues. You might have to wait 45 seconds before you hear the phone ring on the far end. When someone picks up the call, you might hear static, echo, or low volume on the call, making it difficult to hear what the other person is saying. You can still have a conversation, but it becomes a bit more challenging than you might want. It's not fun.

If a fax machine experiences any of these issues, the fax will probably fail. Any issue that prevents the clean transmission of data in a timely manner typically kills a fax transmission.

There are four main reasons international faxes fail, and unfortunately, there isn't a lot you can do to change them. They are:

- **Antiquated network in the destination country:** If you are trying to fax to a company in rural China or a small village in Africa, for example, the network might be old, outdated, and inherently prone to dropping calls. The company might be maxing out its bandwidth, and without enough bandwidth to allow the fax machines to synch up after the call connects, your fax will fail. ***Solution:*** Make multiple attempts at a lower baud rate in order to complete your fax.
- **Receiving fax can't increase baud rate fast enough:** If your fax is transmitting at 64 kbps, and the receiving fax is 20 years old and has a maximum throughput of 1200 baud, there might not been enough time for your fax machine to slow down and the remote fax to speed up before your fax machine loses synch and drops the call. ***Solution:*** Your best bet is to lower the speed of your fax machine and keep trying the call. Eventually, it should complete.
- **Fax machines time out before connection:** Every fax machine has a timer on it that disconnects a call if it's not answered. The factory default on this timer is about 30 seconds, and for domestic calls, that is acceptable. The problem is that international calls take more time to set up. There can be a 30- to 45-second PDD on an international call. If the fax machine you are dialing to picks up on the second ring, it might be a full minute before your fax machine receives a connection. ***Solution:*** The easiest fix is to simply set the *wait for connect timer* on your fax machine to 2 minutes, or at least to 90 seconds.
- **Carrier compression techniques on international calls:** Some underlying carriers work so hard to maximize their profits that they squeeze every call going over their network to tenuous levels. This isn't much of a problem on voice calls, but fax machines need enough bandwidth to allow them to synch up after the call connects. If there is too much latency on the call, the fax machines won't synch up and the call will fail. ***Solution:*** There might be nothing you can do to resolve this problem, except to open up trouble tickets and push your carrier to move the traffic to another underlying carrier.

Chapter 12

Troubleshooting Switched Network Issues

In This Chapter

- Troubleshooting switched outbound calls
- Validating your long-distance carrier
- Calling out over another long-distance carrier
- Using another local carrier for outbound calls
- Troubleshooting switched toll-free calls
- Validating your ring-to number
- Confirming the RespOrg
- Understanding outlying area issues with toll-free numbers
- Troubleshooting international toll-free numbers

Thoroughly troubleshooting any problem before you call your carrier or hardware vendor greatly reduces the time repairing issues takes. You might already have a general indication as to the source of the problem by simply comparing the types of calls affected, but the tests in this chapter allow you to pinpoint the source of the problem.

The tests in this chapter are the same ones your carrier uses, or asks you to try, when you call to open a trouble ticket. Your carrier can quickly bring the troubleshooting process to greater focus when you have already eliminated most of the variables and isolated the source of the problem. This chapter covers switched troubleshooting on both outbound and inbound toll-free calls. You will probably refer to this chapter several times a year, so keep a bookmark here.

If the problem affects different types of calls, it's always best to open your trouble ticket as a switched outbound call. A single problem in the switched network affects all calls that pass through the affected switch or area of your carrier's network, including toll-free calls and calls that originate or terminate on your dedicated circuit. By opening the ticket as anything but a switched outbound issue, you can delay the repair process by temporarily diverting your network's attention to your circuit or your toll-free number.

Troubleshooting an outbound call doesn't require many steps, but each step is very important. After you become comfortable with the entire troubleshooting process, from comparing the call types to using the tests to isolate or bypass the individual variables, you can identify problems in a few seconds.

Doing Background Work Before You Begin Troubleshooting

The key to good troubleshooting is being systematic. That means you have to do some preliminary work and get yourself organized before jumping in and making a bunch of random test calls. Read the following instructions for more information about what to do before you begin troubleshooting.

Determining whether the problem is a provisioning issue

Before you start testing calls, begin troubleshooting the issue by asking yourself whether the feature or problem area ever worked. If so, when did it change and why? These are important questions because the process you follow for fixing *provisioning issues* is different than the process for troubleshooting *trouble issues.* Your carrier's order entry (or provisioning) department handles provisioning issues, whereas the trouble repair people work on a service after it's installed.

If you have problems with a service that never worked, it's worth your while to place a 60-second call to the provisioning team at your carrier to ensure that your order was completed. After you confirm that the order has been completed, you can perform the testing in this chapter and call the customer service department to open a trouble ticket. Alternatively, if you are in doubt and don't want to waste time, you can have the provisioning department research the issue while you simultaneously report it to the customer service department. Simply leave a message for your provisioner and then call customer service to open a trouble ticket. This way, you don't lose any time and you cover all of your bases.

Talking to your staff

If your service has been working fine for over a week before you start having problems, you can begin the troubleshooting process. Well, sort of. Before you get started, follow these steps:

1. **Take a quick survey of the people in your office to find out who else experiences the problem.**
2. **Note the types of calls that seem to be affected the most.**
3. **Seek out individuals whose calls were most affected and ask specific questions:**
    - Were they inbound local calls?
    - Outbound long-distance?
    - Inbound toll-free calls?
    - Outbound toll-free calls?
    - Did they happen at certain times of day?

    By comparing the call types (see Chapter 11 for help), you can get a general idea of where the problem resides.

If the problem is pretty consistent, you will have an easier time troubleshooting it.

Starting the Troubleshooting Process

Use the sections that follow to easily isolate any issue that isn't intermittent, including call-quality issues such as static and echo, as well as call-completion issues such as recordings or fast busy signals. If your issue is intermittent, you might need to repeat the tests 10 or 20 times before you can confirm the results.

When you are having a problem on an outbound local call, it can affect all of your calls because all call types will route into or out of your local carrier. To troubleshoot this type of problem, skip to Step 5 and then open a trouble ticket as necessary based on your results.

Step 1: Redialing the number

If you are having a completion issue (say calls fail to a recording or a fast busy signal) you may have misdialed. Try calling back, even if you dial the number all the time and are *sure* you couldn't have misdialed. It's common to transpose a digit or press an incorrect key. If you confirm that you misdialed, you can go on about your day.

If pressing the Redial button on your phone completes your call, you should be aware of the fact that there might have been an intermittent issue you stumbled across, or an outage that just cleared up on its own. If the call still fails or has quality issues when you redial it, move to Step 2.

The key to doing *bad* troubleshooting is to make assumptions. Even the most basic assumptions (of course I dialed right!) can lead to wasted time.

If you have call quality issues, write down the information about all of your call examples (see Chapter 11 for the specifics of what you need) and try the number again. Call quality issues only affect calls that pass over the affected circuits or through the affected hardware. If the problem is deep enough into your carrier's network, the static or echo may be intermittent; if your call is sent over a different circuit to reach its destination, you might not have any problems. If the problem affects all of your calls, proceed to Step 2 to continue troubleshooting.

If you can't finish dialing the entire phone number before the call fails, or if you have a line-quality issue before you finish dialing the phone number, the problem is either with your hardware or your local carrier. Your call won't make it to your long-distance carrier's network until after the last digit is dialed and your local carrier identifies it as a long-distance call. If you have static on the line after you dial three or four digits, but before you finish dialing the phone number, the issue is probably with your local carrier. If you have a PBX phone system that performs any type of least-cost routing, your local carrier won't see the call until after you enter the last digit and your PBX decides where to route it. In this case, any static or issues that occur before you finish the dialing sequence are the result of hardware problems within your phone system. There might be a bad port or line in the phone system that is causing it. Because the tests indicate it's your hardware, you need to dispatch your hardware vendor to fix it.

Step 2: Validating your long-distance carrier

If you think the problem is with your long-distance carrier, you need to confirm that your calls are actually going to the carrier. It's possible that some other long-distance carrier accidentally changed your phone line over to its network, and then blocked your access to its network after realizing the mistake. In this case, the reason your long-distance calls aren't completing is because your calls aren't running over the correct network. Of course, that's an assumption — you don't know for sure until you check. You need to test part of the call that moves from your local carrier to the long-distance carrier (see Figure 12-1).

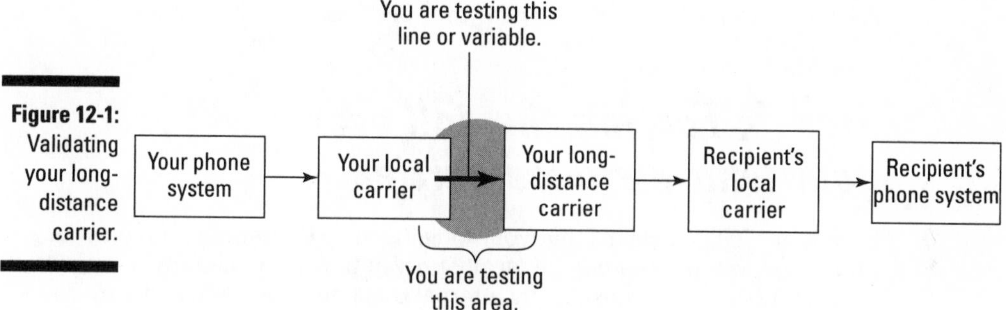

Figure 12-1: Validating your long-distance carrier.

The industry standard test for validating your long-distance carrier is called a *700 test*. The 700 test got its name because you dial 1-700-555-4141 from your touch-tone phone (rotary dial phones may not have access to the service). The 700 test is a free local call and is routed by your local carrier to a recording that identifies your interLATA long-distance carrier. If your long-distance carrier is AT&T, for example, the recording says, "Thank you for using AT&T." This test is almost always accurate, so investigate anything that doesn't sound correct.

Performing a 700 test now, before you have a problem, can help you avoid possible confusion in the future. Your phone bill might come from one carrier, but if the carrier uses a subsidiary to provide your service, the 700 test could lead to a confusing message. The number of *switchless resellers* continues to increase; additionally, carriers are often bought, sold, and renamed, so knowing who your carrier is can be a lot more confusing than it seems. For example, maybe your local carrier doesn't know that IXC Communications purchased Cincinnati Bell and is now Broadwing, which is why you don't hear the new Broadwing recording that you *should* hear. If your 700 test lists a carrier you don't recognize, call your long-distance provider for validation. If you are wondering what a switchless reseller is, I suggest checking out Chapter 1 for a crash course.

Some competing local exchange carriers (commonly called *CLECs*), such as PAETEC and Mpower, might not provide the 700 testing feature, but might instead send you to an automated voice reciting the phone number you are dialing from. In this case, the only way to determine where your local carrier is routing your long-distance calls is by calling your local carrier.

The 700 test only validates the long-distance carrier for an individual phone line. If you want to check multiple phone lines, you have to test them all manually, one by one.

The 700 test only validates the *primary interexchange carrier* (PIC) your local carrier has listed for your *inter*LATA long-distance traffic. Some local carriers offer a 700 test for your *intra*LATA PIC of 1-700-your area code-4141, but generally, you can only confirm the long-distance carrier for your interLATA calling. If you aren't sure what the difference is between intraLATA and interLATA calls, please check out Chapter 3 for the specifics.

Step 3: Forcing the call over your long-distance carrier

Even if the 700 test claims that your long-distance carrier is set up on your phone line, you haven't unequivocally proven that your calls are being routed to there. Local carriers use many internal tracking databases, and only one of them handles 700 tests. As I said in the previous section, 700 tests are *usually* correct, but a totally different system deeper in the computer switch (a switch that is *always* accurate) actually routes all of your long-distance calls.

Being both persistent and polite, ask the customer service representative at your local carrier to do a conference call, or simply ask to have a switch technician validate your long-distance carrier. If the switch identifies the long-distance carrier as MCI, but the 700 test says your long-distance carrier is Sprint, the 700 recording is wrong — your calls are transmitted across the MCI network. The switch is always right, because the switch is the piece of hardware that actually sends your call to your long-distance carrier. Figure 12-2 shows that the area tested is still within the realm of the local carrier (refer to Figure 12-1). The difference is that in this test the focus is on a section of your local carrier's network a little farther downstream.

You may have noticed something peculiar about Figures 12-1 and 12-2. The pictures are exactly the same, it's true. Both Step 2 and Step 3 involve validating the routing by your local carrier. However, Step 2 (represented in Figure 12-1) tests to ensure that the call is being sent to the correct carrier; the second test (Step 3, represented in Figure 12-2) affects the circuits used to reach that local carrier. I know, I know, I'm splitting hairs, but the point is to help you identify that you need to talk to your local carrier in each of these instances, even though at first blush it looks like you're dealing with a long-distance problem.

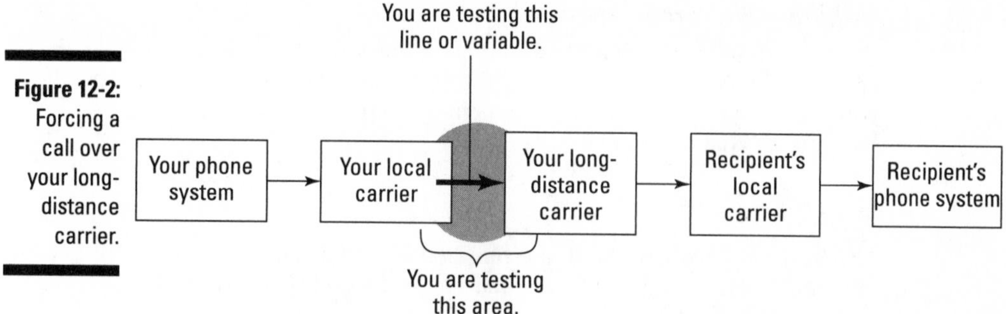

Figure 12-2: Forcing a call over your long-distance carrier.

You do have the power to force calls onto your long-distance carrier. All it takes is the *dial-around code* for your long-distance carrier and the capability to dial seven more digits. Every long-distance carrier has a dial-around code, and if you have any switched phone lines, you should know the one that belongs to your carrier. When you ask for the code, the carrier gives you a number that begins with 10-10 (pronounced *ten-ten*) and ends in three more digits. For example, the dial-around code for AT&T is 10-10-288; for MCI, it's 10-10-222, and for Sprint it's 10-10-333. If you want to find out more about dial-around codes, you can meander over to Chapter 7 and read the section on *casual dialing*.

Performing the 700 test validates only one part of your local carrier's network, the section that identifies the carrier to receive your long-distance calls. The other part of the local carrier's network that you need to be concerned about is how the local carrier actually routes your calls to your long-distance carrier after the local carrier's network determines the correct route to the correct long-distance carrier. If the circuits making up the route to your long-distance carrier aren't functioning correctly, your call fails, even though the 700 test has correctly identified the long-distance carrier. *Forcing a call* over your long-distance network bypasses the first step and can result in your call taking a different path to your long-distance network.

How dial-around codes work

A dial-around code tells your local carrier to immediately route your call onto the long-distance network associated with that particular code. Dialing a phone number with one of these 10-10 codes bypasses the local carrier process that determines whether your call is local or long distance, and identifies which carrier is assigned to your phone line for long distance. Your local carrier simply throws your call to the network attributed to that dial-around code.

Keep the 10-10 dial-around code for your carrier someplace where you can't lose it. If you have to print it on an adhesive label and stick it to the underside of your phone, or on the wall next to your phone system, do it.

Making a dial-around test call

After you determine the dial-around code for your long-distance carrier, try your problem call again by dialing the 10-10 code and the phone number. If your carrier is Sprint and you are having problems dialing someone in Milwaukee, Wisconsin, dial

```
10-10-333 + 1 + 414+ the phone number
```

Always dial `1` between the dial-around code for your carrier and the phone number that is giving you problems. The call won't work without it.

Some phone systems purposely block your ability to use dial-around codes. If the call fails before you finish dialing all the digits for the phone number, your phone system is probably the cause. Either call your hardware vendor and find out whether you can use a bypass code, or locate a phone line that doesn't go through your phone system and try the 10-10 code from there. Fax and modem lines are frequently left out of phone systems, so you may be able to use them for testing. If you do receive a bypass code for your phone system, it will probably be complex, convoluted, and might require several steps before you reach a dial tone from your local carrier.

If the call still fails in Step 3

If the call still fails in the same way it did before, the good news is that your local carrier is sending the call to your long-distance carrier just as it should. The bad news is that you still have the problem, so proceed to Step 4.

If Step 3 resolves the problem

If the call doesn't fail when you dial using the dial-around code for your carrier, but it does fail when you call out directly, your local carrier is the most likely source of the problem. If you have a standard local carrier like Bell Atlantic or Verizon, you can just dial 611 on your phone and follow the voice prompts to be directed to the repair department. If you have a *competing local exchange carrier* (CLEC) such as Mpower, PAETEC, or US LEC, you have to find their phone number. If you are having problems finding their phone number, check out the section about finding and contacting your local carrier in Chapter 7.

International calls generally have a *post dial delay* (*PDD*) of 15 to 30 seconds. The PDD is the time after you finish dialing the phone number but before you hear the phone ring on the other side. You can generally reduce the PDD by as much as 12 seconds by using your carrier's dial-around code. If you have international fax problems caused by the PDD, this might solve them.

Keeping your local carrier from passing the buck

When you tell your local carrier that you have problems dialing a long-distance number, the customer service rep will quickly try to end the conversation and direct you to your long-distance provider. Typically, the local carrier's rep is right to direct you to the long-distance carrier.

You can easily clarify the situation by telling the person that you dialed the number with the dial-around code and it completed, but that when you dial it without the code, the call fails. The customer service rep will then know that you have tested the issue and you know what you are talking about. As long as you continue to press this point, the rep will open a ticket for you and repair the issue. If you want to really impress the rep, tell him or her you completed a 700 test before you used the dial-around code.

Step 4: Dialing over another long-distance carrier

If you suspect that you have a long-distance issue, Step 4 enables you to redeem or condemn the long-distance carrier. Up until this point, you have simply validated that the call is reaching the correct long-distance network; now, you can remove your long-distance network from the call entirely and replace it with another carrier. You accomplish this task the same way you forced the call onto your long-distance carrier's network (see Step 3, earlier in this chapter); this time, however, you use the dial-around code of another long-distance carrier. The section of a call isolated by this step is shown in Figure 12-3.

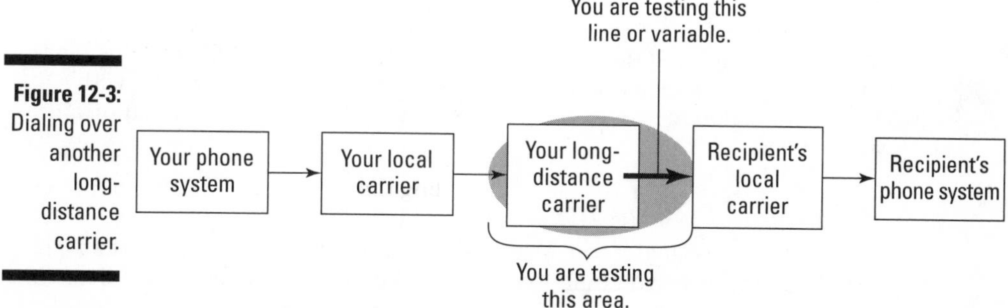

Figure 12-3: Dialing over another long-distance carrier.

Using 10-10 codes for international issues

There is strange magic in the world of telecom that doesn't make much sense to the logically minded. You use the dial-around code to reach the same entrance point into your long-distance carrier that your local carrier uses. For some reason, using the dial-around code seems to speed the connection of international calls.

You can choose among the following carriers:

- **10-10-288:** AT&T
- **10-10-222:** MCI
- **10-10-333:** Sprint

If you want to validate your carrier by sending your call over another network, try your calls over the strongest networks you can find. Sprint and AT&T have outstanding networks, as does MCI. If your call fails over Sprint, make two more test calls (one using MCI and the other using AT&T), just to confirm the local carrier that owns the phone number you are dialing is actually the problem and not your long-distance carrier's network.

Making a dial-around test call with a different carrier

If you're a Sprint user and you want to test a number in Milwaukee, Wisconsin, to determine whether the call might work on the AT&T network (thereby proving that Sprint carrier is the source of your problem), dial

```
10-10-288+ 1 + 414 + the phone number
```

If the call completes when you are trying it over one of these carriers, be aware of the fact that unless you are set up for casual dialing over this long-distance carrier's network, you might be charged $0.55 per minute, with a $3.75 connection fee, or more. By all means, make that call if it's important; just make it quick.

If the call still fails in Step 4

If you try to use one or more other long-distance carriers and the calls still fail, I can guarantee you that your long-distance carrier isn't the problem. You still have three other areas left to troubleshoot, and Step 5 can possibly eliminate two of them.

If Step 4 allows you to complete your call

Completing your call over another long-distance carrier confirms that the source of the failure is within your long-distance network. Call your carrier and list all the tests you have tried, and their results. The representative you talk to might have to follow a troubleshooting script with prompts to ask for some of this information, but be sure to report your specific test information:

- The results of your 700 test
- The results of dialing out, using the 10-10 dial-around code
- The date, time, origination number, termination number, and call treatment of your failed call, as well as the call you made with the 10-10 dial-around code
- The name of the carrier whose network you used to complete the call

Take notes, and be sure to write down the trouble ticket number and contact name and extension for the person you talk to. Follow up with the carrier every two hours to see how the situation is coming along.

Using Step 4 for international issues

Dialing out over another carrier is very helpful in troubleshooting international issues. If you can tell your carrier's rep that the call is completing over AT&T and Sprint, he or she won't worry that the problem originates in a foreign country.

Some long-distance carriers have triggers in their networks to prevent fraud; frequently, those triggers involve international calls. I have never encountered a fraud block dialing out over Sprint, MCI, or AT&T, but if you use Qwest (its dial-around code is 10-10-432) to make an international test call, your call will probably be blocked from the Qwest network. The only way to remove the fraud block is to open an account with Qwest and sign up the specific phone number that is blocked. That's a lot of work just to do a troubleshooting test that your long-distance carrier will probably do anyway.

The long-distance carrier whose 10-10 code you use will charge you quite a bit of money for completed calls. If you have to make an international call over another carrier, keep it short, or expect to pay the consequences.

Step 5: Dialing from another local carrier

The tests in the previous sections validate some of the handoff between your local and long-distance carrier, as well as the entire path of the call through your long-distance network. Although a handful of potential issues remain, only two of them are issues that you can take control of: they are your hardware and your local carrier. Unless you have been using the bypass code for your hardware to make the 10-10 calls, or you don't have a phone system, your local carrier is still a possible source of the failure. By using a different local carrier, you can confirm or disprove whether your local carrier is routing your calls correctly. Figure 12-4 identifies the section of the call we are isolating during Step 5 of testing.

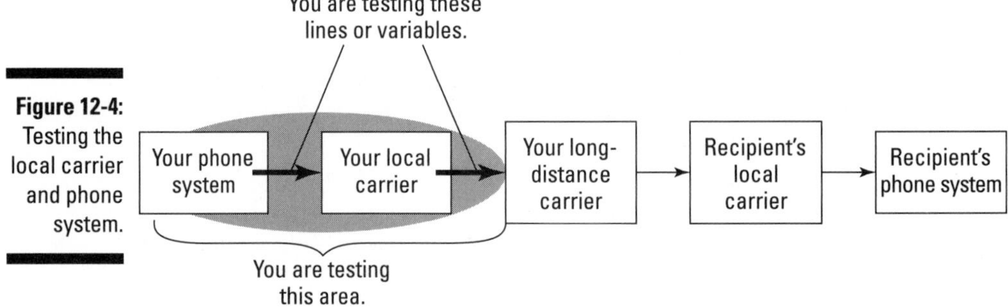

Figure 12-4: Testing the local carrier and phone system.

Making the test call over another local carrier

You can make this test as complex or a simple as you want. If you want to make it difficult, ask someone in another state to make the test call for you. If you want to make it easy, just grab a cellphone and dial the phone number you want to test. Technically, wireless providers are competing local exchange carriers (CLECs) and provide the same service as your local carrier. Unless you somehow have the same local carrier for your cellphone and your land lines, this test gives you a lot of information.

If Step 5 allows you to complete your call

If the call completes from your cellphone, your local carrier or your phone system is the cause of the problem. If you don't have a phone system, or if you were bypassing all the routing within it for your test calls, the issue is probably with your local carrier. Even if you haven't bypassed your phone system and you haven't ruled it out as a source for the problem, open a trouble ticket with your local carrier anyway, providing the following information:

- The results of your 700 test
- The results of your dial-around tests, using your carrier's code, and using the codes of other networks
- The time of your failed calls, the origination phone number, termination phone number, and call treatment
- The fact that you can complete the call from a telephone outside your phone system

If your local carrier doesn't find any problems on your phone line, can access a dial tone from its central office switch, and can complete a call, the problem lies with your phone system. It's time to call your hardware vendor.

Keep the trouble ticket open with your local carrier so that when your hardware vendor arrives, he or she can conduct a head-to-head test with the carrier. In a head-to-head test, the hardware vendor can make test calls, and your local carrier can watch each digit come across the network as the technician dials them. Between these two people, your problem will be fixed.

If your call still fails in Step 5

If you have the same failure when you dial from a different local carrier, you know that nothing from your phone to the end of your long-distance network is causing the problem. In this case, there are only two potential trouble areas left: the local carrier that provides a dial tone to the phone number you are dialing and the phone system that receives your call. If you know that the person you are calling doesn't have a phone system like a PBX or Key system handling their calls and providing features like voicemail (he or she uses plain old telephone service), the problem is most likely with his or her local carrier. If you aren't sure what constitutes a phone system, you can find out more in Chapter 2. Consider any phone system other than a single-line phone plugged into the wall (or two tin cans connected by a piece of twine) as a phone system.

This is when you have to decide exactly how much you want to make this call. If it's not that important, try calling back in two or three days. The local carrier on the other end will eventually identify and repair the problem.

Getting Switched Toll-Free Troubleshooting Basics

Toll-free troubleshooting is slightly more complex than standard outbound troubleshooting because handling of toll-free numbers can vary by state, LATA, area code, or by area code and the next three digits in the phone number. There are also the dual concerns of who the RespOrg is and who carries the traffic. Finally, problems with toll-free numbers might not initiate from the toll-free number at all, but can be a problem with the regular phone number it rings into, or just a systemic network issue affecting all calls that run through the point of failure.

Before you begin troubleshooting your toll-free number, it's important to validate the toll-free number's basic information.

- **The toll-free number or feature:** Has it been working in the past? If you are in the process of moving your toll-free number from one carrier to another, or adding a new service, you can have any number of provisioning problems that prevent the service from being activated. Provisioning issues are handled by your carrier's order-entry department; see Chapter 9.
- **The ring-to number:** If you ordered the toll-free number to ring to your fax machine rather than the main phone line for your customer service department, that might be why everyone who calls receives an unpleasant squeal in the ear. Check the phone number to which the toll-free number points. If your order is wrong, possibly that is the source of the problem. If nothing else, you need to know the ring-to number of your toll-free number for the testing process.
- **The area of coverage:** If you ordered only U.S. 48 states coverage and calls are from Alaska and Hawaii fail, the source of your problem lies with the fact that you didn't request access from those areas. Coverage on toll-free numbers can be delivered in many different levels, and your available options are dependent on what your carrier provides. Some carriers offer many options that include or exclude Canada, Alaska, Hawaii, Guam, U.S. Virgin Islands, and/or Puerto Rico, whereas other carriers give access to all these options by default.

You can troubleshoot any issues, either quality- or completion-related, with the steps in the following sections. If you have quality issues, you need to gather as many call examples as possible, both affected calls and clean calls, so that your carrier can isolate the issue (see "Talking to your staff," earlier in this chapter). Completion issues are much easier to find, because they have a definite failure point that technicians can duplicate and find.

Step 1: Dialing the number yourself

When a customer calls in to your direct phone number to report that your toll-free number is failing, first make sure you find out the following information from your customer before you hang up the phone:

- The phone number and location from which your customer is calling
- What time the customer's call failed
- The call treatment

Then try to dial the number yourself. It's more common for a toll-free number to fail from all locations than from only a specific geographic area, such as a LATA or a state. If your call fails when you dial the toll-free number, all inbound calls are probably failing, regardless of the location of the caller.

If your call to your toll-free number completes, there is always the possibility that the problem is restricted to the area your customer is calling from, which is why you need to know the phone number and location of the caller. You can ask the customer to help you troubleshoot the number, or you can call your toll-free carrier to open a trouble ticket.

You should also ask your customer to place you on hold so you can try to call the toll-free number again. The customer could have dialed the number incorrectly. If the customer can complete the call, there's a very strong chance that he or she simply misdialed the number the first time. If the call fails again, there's an equally strong chance that the problem is of a geographical nature. Until you receive additional reports of problems from your customers, or you get feedback from your carrier, you won't know how large the affected area is.

Write down all of your call examples and call your carrier. When you open your trouble ticket, be sure to write down the date and time, as well as the customer service rep's name. After the ticket is open, follow up on it every two to four hours until the issue is resolved.

If the call fails when you dial your own toll-free number, the outage might be affecting everyone calling, but at least you have the ability to troubleshoot the issue. In this case, clear off a page in your notebook for test calls and proceed to Step 2.

Always try to find someone in an affected area to retest to your toll-free number. Long-standing customers or vendors will generally be more than happy to spend five minutes making test calls. If your toll-free number always completes when you call it, you don't assume the issue is fixed until you receive a call from the area that was previously blocked.

Step 2: Dialing the ring-to number locally

Calling the ring-to number directly can identify whether the problem is within the toll-free routing portion of the call or is caused by local carrier problems reaching your regular phone line. Figure 12-5 shows the area of your local carrier's network that we are validating in this step.

If you are calling from the same office that receives the toll-free number, you place a local call to validate the ring-to number. Local calls pass through your phone system and your local carrier, so if you can complete this call, you know these two portions of the system are working.

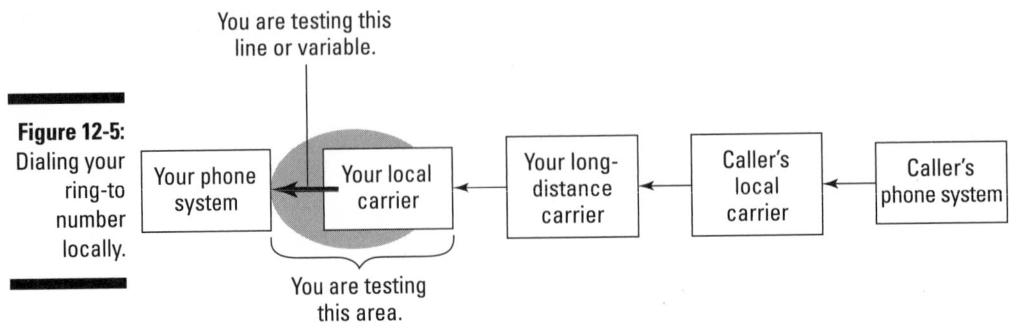

Figure 12-5: Dialing your ring-to number locally.

If the local call to the ring-to number fails in Step 2

If your local call fails, the issue has little to do with the toll-free number. You can't complete a toll-free call if your local carrier can't complete their portion of the inbound call. You can't be positive at this time whether the problem is

- Caused by your local carrier being unable to route calls to your office
- Within your phone system

If you don't have a phone system, the only possible source of the issue is your local carrier.

In this case, collect all the information about your inbound local call examples, and dial 611 to reach your local carrier to open a trouble ticket. Write down the trouble ticket number, whom you spoke with, the time and date, and then you should follow up periodically until the problem is resolved.

Don't mention anything about your toll-free number to your local carrier when you open your trouble ticket. If you try to explain that you have a toll-free number that isn't working and that you have been troubleshooting it, the local carrier will identify the number's RespOrg and direct you to call your long-distance carrier. Instead, open the trouble ticket on a call example that shows that you can't receive a local inbound call. That way, the carrier knows the issue exists within its own network. As long as your local carrier isn't confused by stories of toll-free numbers, the issue will repaired in short order.

If your call completes in Step 2

If your call fails when you dial the toll-free number and completes when you dial the ring-to number, you can rest assured that neither your phone system nor your local carrier are the source of the problem. In order to resolve the problem, move on to Step 3.

Step 3: Dialing the ring-to number through your long-distance carrier

Your next step is to determine whether the issue is legitimately in the toll-free handling of the call, or if it's simply a larger network issue within your long-distance carrier. In order to truly validate how your long-distance carrier is sending calls to your phone number, you need to have a call placed over the long-distance carrier's network to your ring-to number. The specific area of the call you're focusing on in Step 3 is shown in Figure 12-6.

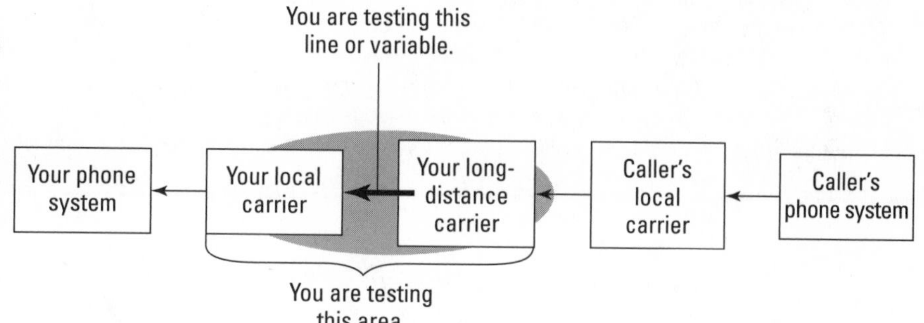

Figure 12-6: Dialing your ring-to number through your long-distance carrier.

If your local carrier provides your toll-free number, the carrier has a complementary carrier that provides the long-distance service. In this case, you need to know the 10-10 dial-around code for your long-distance carrier so you can make the test call.

Test calling the ring-to number on your long-distance carrier's network

Even though calling from one line in your office to another line in the same office is a local call, you can force the call to be routed over your long-distance carrier and back in through your local carrier. If you preface your call with the 10-10 dial-around code of your long-distance carrier, the call will be sent from your local carrier to your long-distance carrier, and then forwarded back to your local carrier to ring to your office. It's a short loop, but even a short loop can indicate the general health of your long-distance network. If your long-distance carrier is Sprint, dial your phone number like this:

```
10-10-333 + 1 + area code + your phone number
```

Use 10-10-288 if your carrier is AT&T; use 10-10-222 is your carrier is MCI.

Understanding inbound routing

The most important piece of information in the world of telecom isn't the physical location of the phone, but the physical location of the central office that provides your dial tone from the local carrier. Every call you receive is sent to that central office and forwarded to you.

Periodically, your local carrier might decide that it has too many calls running through one central office, so it moves a group of phone numbers to a different central office in the area. The local carrier then updates the national Local Exchange Routing Guide, or LERG database (which I explain in Chapter 3) to inform every telecom company about the change. However, if your long-distance carrier is slow to implement the change, it might continue to send calls to the old central office, causing them to fail.

If your carrier is handling your toll-free calls, those calls will also fail. The repair can take as little as 15 minutes if your carrier can identify the issue, or it can drag on for days. If you are aware of any changes your local carrier has made to your service, you should let your carrier know, because it can reduce your repair time.

If the test call completes in Step 3

If your call to the ring-to number, using your long-distance carrier, completes without any problem, you have validated that at least a portion of the switched network is working. Carriers commonly route toll-free calls differently than direct-dial calls. Regardless, you must accomplish your next level of research with the help of the customer service reps at your long-distance carrier, so skip the next test and jump to Step 5.

If the test call in Step 3 fails

If your call to the ring-to number over your long-distance carrier fails, you might have an issue with your long-distance carrier's switched network. The good news is that your toll-free number isn't the source of the problem. The bad news is that your long-distance carrier hasn't updated the latest LERG database and is routing your calls to the incorrect central office (CO) of your local carrier (if you are having call completion issues) or the carrier is sending the call over a defective route path (if you have quality issues). What you need to determine next is whether any long-distance carrier can complete a phone call to you. To find out, move on to Step 4.

Step 4: Dialing the ring-to number over another carrier

So far, if you've been following all the preceding steps, you have isolated the problem with your toll-free number to the path your call is taking through your long-distance carrier to your ring-to number. If your local carrier is servicing your phone number out of a different central office, it's possible that only your local carrier knows where to find you. Figure 12-7 shows the area of a toll-free call you are isolating by performing the test in Step 4.

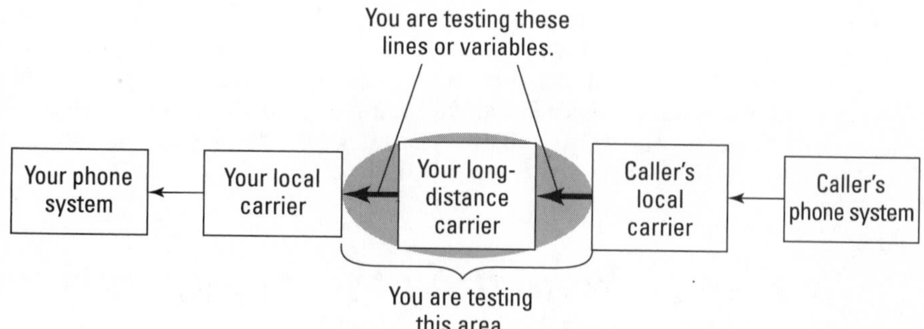

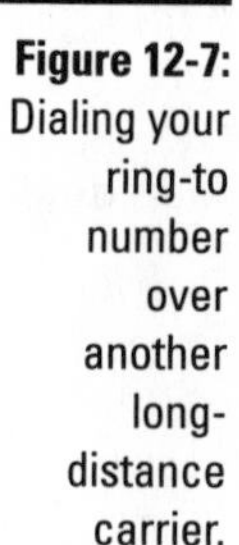
Figure 12-7: Dialing your ring-to number over another long-distance carrier.

Use a different long-distance carrier to validate the inbound route path. That is, if you use Sprint, use the dial-around number for AT&T. In this case, the test call would look like this:

```
10-10-288 + 1 + area code + your number
```

If the call completes, AT&T charges you for the call, at any per-minute rate it chooses. If the call doesn't connect, you won't be charged for the call. If your test call completes, keep the conversation short to prevent being shocked when your invoice arrives.

If the call in Step 4 fails

Try a few other long-distance carriers, such as MCI (10-10-222) and Qwest (10-10-432). If these calls fail, as well, your local carrier has hidden the inbound path to your phone from all long-distance carriers. In this case, take all the call examples for the multiple carriers you have tried and dial 611 to open a trouble ticket with your local carrier. Just as with all trouble tickets, you need to write down the trouble ticket number on the paper alongside your call examples and follow up every few hours.

If the call in Step 4 completes

If the test call to your ring-to number completes over another carrier, your testing is over. This test only changes the long-distance carrier used, and every other variable remains the same. Because the results changed when you changed the long-distance carrier, the problem has to be with that carrier.

When speaking to your carrier, open your trouble ticket as an outbound call issue and *don't* mention your toll-free number if you can avoid it. Instead of telling the carrier that someone unsuccessfully tried to dial in on your toll-free number, tell the carrier that someone with the same long-distance carrier tried to call you and they failed, so you did a test call that failed. If you begin talking about your toll-free number, the carrier might open the trouble ticket on the toll-free number instead, which will delay everything.

When you open the trouble ticket, be sure to tell the technician that you forced the call from your office to the other phone number in your office by dialing the 10-10 dial-around code. Then, as usual, write down the trouble ticket number, the name of the person you spoke with, and the time you called; also make note of any additional testing that was attempted and changes made to the toll-free number or the network. Then set up a schedule to call back every few hours for updates.

Step 5: Validating the ring-to number and RespOrg

If you end up at this step, you have eliminated all the other steps that can possibly be causing this problem, except for the processes and mechanisms that make toll-free routing unique to outbound calling. You should also log some failed call examples to your toll-free number and some completed call examples to your ring-to number.

After you reach a customer service representative, you need to validate two more pieces of information before you open a trouble ticket:

- **Confirm the ring-to number.** This number might have been updated or input incorrectly. If your area code is 206 and the ring-to number is going to area code 602, you have found your problem. In about 15 minutes, your carrier can point your toll-free number where it needs to go and it will be working fine.
- **Confirm the RespOrg.** It's uncommon for a working number to suddenly be moved from one long-distance carrier to another, but it does happen. If your toll-free number somehow migrated to another carrier by accident, someone at the new carrier might have later realized it didn't belong to that carrier and blocked it. If the RespOrg isn't your carrier, you have to begin the process to move it back. Please take two aspirin and go to the section in Chapter 9 about activation issues. Unfortunately, your problem is considered a provisioning issue and not a trouble reporting issue. This means that carriers deal with it with less urgency; the timeline to repair can be 7 to 10 days, as opposed to 15 minutes to 3 hours.

This is the one time when screaming might pay off, as long as you are screaming at people who can actually help you, and that they are people you will never speak to again. I am mainly talking about the conversation you will have with the carrier that somehow is now in RespOrg control of your toll-free number. If you add emotional urgency to your discussion with the representative you speak to, he or she might not hide behind the normal 7- to 10-day time frame for migrations. Before you have this emotionally charged conversation, however, chat with your carrier first to make sure it wasn't the carrier's fault for letting go of the number. Also, at this time, you might choose to NASC your toll-free number back if the carriers involved are moving slower than you require (see Chapter 9).

If the ring-to number and the RespOrg match up fine, proceed with the trouble ticket. Make note of the trouble ticket number, the person you spoke with, the time and date you chatted (in case the problem takes more than one day to repair), and some brief notes that you have checked RespOrg, and the ring-to-number. Then, set your alarm clock to ring every two or three hours so that you can follow up and keep the ticket moving forward.

Troubleshooting Toll-Free Issues from Canada, Alaska, and Hawaii

Troubleshooting toll-free issues is difficult enough, but it gets even worse when the problematic calls are unique to areas outside of the lower 48 U.S. states. Your first order of business is to confirm that the toll-free number covers the areas in question. When your business was started, maybe Canada or Hawaii weren't part of your target market. As your company grows, you find a demographic in one of these areas that is a perfect fit with your company's business plan. Contact your long-distance carrier to validate the area of coverage on your toll-free numbers and open up the coverage to the area as soon as you identify the need.

The process of adding coverage to your toll-free service takes as long as three to five days, so it is best to deal with it immediately, before you have a squad of executives at a hotel in Victoria, British Columbia, wondering why they can't call back into the office. If the execs are on a short business trip, you might get the toll-free numbers working from Canada — just about the time they land on their flight home.

This advice may seem simple, but don't assume that your service coverage includes calls outside the continental U.S. If you first confirm that your toll-free numbers have been set up for calls from these outlying areas, you can make things happen much quicker.

To simplify the troubleshooting examples here, I use the scenario that one of your toll-free numbers isn't accepting calls from Canada. The troubleshooting scenario is roughly the same for any of the other areas, including Alaska, Hawaii, the U.S. Virgin Islands, and Puerto Rico.

Getting a sample troubleshooting scenario for outlying areas

Maybe your toll-free number was working before, or maybe it was set up to receive calls from Canada a while ago but you never verified this access. Whatever the issue, calls from Canada are not reaching your toll-free number. Almost all the issues concerning outlying areas have the same profile, regardless of location:

- The toll-free number being dialed is supposed to be open for calls from the area.
- The toll-free number works from everywhere in the lower 48 states.
- When calling the toll-free number from Canada, callers hear a recording such as, "The number you have dialed is unavailable from your calling area."
- The person having the problem with the number is unavailable for retesting.
- You have limited or no information about the dates, times, and phone numbers of the calls that are failing from Canada. Your executives at the trade show don't care about the details; they just want the darn thing fixed.

If you look at the situation from a troubleshooting standpoint, it might seem like you're in a hopeless situation. You have an irritable person in a remote location on a tight timeline who can't give you any details about the failed call and who is essentially unavailable for scheduling any future testing. On top of that, the person is probably in a position of power at the company and your life may be affected in a direct (and negative) manner if you can't fix the problem. Your carrier wants specific call examples and call treatments, and has questions in need of answers. After your customer service representative realizes that you have none of this information, that will be the end of the conversation.

Most long-distance carriers don't directly handle coverage on their toll-free numbers from outlying areas such as Alaska, Canada, the U.S. Virgin Islands, and Puerto Rico. As a general rule, they use an *underlying carrier* who has already made all the connections in those areas. This means that there are two systems that need to be in synch: Your carrier must list the toll-free number as having access from the outlying area, and the underlying carrier for that outlying area must have your toll-free number listed for coverage in the area and on the account for your long-distance carrier. If the underlying carrier in Canada doesn't have your number listed for service with your long-distance carrier, calls to your toll-free number will fail in Canada and never reach your carrier.

Fabricating a call example

What I am about to propose is *top secret.* The likelihood that there is one area in Canada where the underlying carrier isn't processing your toll-free calls is quite slim. It's much more common for the underlying carrier in Canada to somehow drop your number from its records and now calls aren't being sent from anywhere in Canada to your long-distance carrier in the U.S. Fixing this specific problem takes a few hours. Your carrier simply needs to send an order to the underlying carrier in Canada to place the number back online.

After the underlying carrier updates its system, the situation is resolved. The only problem with this scenario is that if you try to ask someone in the order entry department at your long-distance carrier to resend the order for Canada coverage back to their underlying carrier, the carrier rep will see that it's already set up for Canada coverage. The rep might be unable, or unwilling, to order service that already appears to be in place. In the eyes of the rep, the job is done. At this point, the rep refers you to speak to a customer service agent to open a trouble ticket. This is not what you need! In order to have your carrier do what needs to be done, you first need to fabricate a call example.

Don't fabricate call examples under any other circumstances, because doing so will only frustrate you and your carrier, delaying the repair of your problem! The only reason this strategy is acceptable is because:

- **The carrier needs to get its records up to date.** If you suspect the underlying carrier simply has to update its records to bring the toll-free number on to the account for your long-distance carrier, fabricate away.
- **You can't get a call example any other way.** If the underlying carrier isn't routing the calls to your long-distance carrier, your long-distance carrier won't find a valid call example. If the call is stopped before it reaches your long-distance carrier's network, they have no way to trace it, which is the conclusion you are leading them to by making up a fake call example.
- **It's standard procedure for most long-distance carriers to resend the order for coverage to the underlying carrier in this instance.** If there is a trouble ticket on a toll-free number that can't complete from an outlying area, and the carrier can't find the call example, the technicians will suspect the underlying carrier.

If you're ready to call your long-distance carrier and give the person you speak with a falsified call example, you need to make the fake calls sound as legitimate as possible. Your fictitious call example should look something like this:

- **Say that the time of call is about 90 minutes ago.** Give a distinct time, like 11:27 a.m. or 1:35 p.m. The more precise you are, the more likely the call example will go through unquestioned.

- **Say that the call treatment is "The number you are dialing is unavailable from your calling area."** Choose this recorded message to lead the technician to the conclusion that the order to the underlying carrier needs to be resent. Any other call treatment you give will focus attention elsewhere.
- **Use a plausible origination phone number.** Whether you make up the number or use one that is frequently dialed, this is the most crucial part of your fictitious call example. Your carrier is going to search for calls from that phone number, so if you give a completely bogus phone number, you will hear back two hours later that your origination number is invalid, and you will have lost time. You must have a legitimate origination phone number.

If you don't have any reference number in Canada to use as an origination number, you need to find one. The easiest way to pull up a valid phone number in Canada, Alaska, Hawaii, or any outlying area, is to search the Internet. If you know the person is in Alberta, Canada, and can't reach your toll-free number, just search the Web for *alberta canada museums* or *alberta canada university.* Eventually, you will find a home page that displays a local number for either the museum gift shop or the student housing coordinator of the university. It doesn't matter who the number belongs to, as long as it's a valid number.

After you have the phone number, write it in your call example, and dial your long-distance carrier. If you want to add some spice to the story, you can always tell the carrier that your CEO is the one who dialed from this number. Say, "If you could move this along, I may be able to keep my job for a little longer."

In a few hours, your customer service technician calls back to tell you that he or she couldn't find your call example, but that the order has been resent to the underlying carrier. In an hour or so after that, you can retest the toll-free number from Canada.

At this time, you need to ask the executive in Alberta to try the call again. If it fails again, tell the executive you need the origination phone number, the time of the call, and the exact recording or call treatment. In the vast majority of the cases, the fake call example solves this type of problem. If it doesn't, you need to gather legitimate information and push forward. As long as you take good notes, follow up on the ticket, and have someone who can work with you to make test calls in the affected area, there is no problem that can't be resolved.

Troubleshooting International Toll-Free Issues

Every country has its own version of toll-free numbers. Many countries use 0800, whereas others use a variant of it that uses one 8, along with the country's prefix. Some countries have a unique prefix; Japan uses 00531-12 as its area code, for example. The prefix might also change, based upon the local carrier or geographic region from which the call is being dialed. For example, calling a toll-free number from Northern China requires you to dial 10-800-120 + the local phone number, but if you dial from Southern China, you need to dial the number as 10-800-712-XXXX. International toll-free numbers are unique creatures in the world of telecom with a double layer of toll-free numbers built within them, as shown in Figure 12-8.

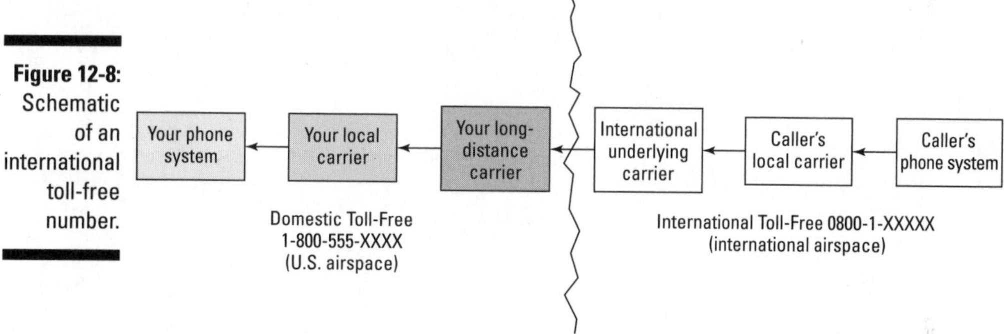

Figure 12-8: Schematic of an international toll-free number.

International toll-free service (*ITFS*) is actually made up of two toll-free numbers. The first number is the toll-free number in the foreign country. This is the number that you dial in England, Tanzania, or some other country that is recognized by their national phone system as being toll free. The second toll-free number is here in the U.S. Its job is to receive calls from the foreign country and route them to your phone line here in America. The differences in troubleshooting an international toll-free number when compared to a domestic toll-free number are few:

- You can't troubleshoot these numbers yourself.
- It generally takes twice as long to resolve the trouble ticket.

- You need to provide the call example with an origination number that looks completely foreign to you. It might seem to have too many or too few digits, but you will have to trust what the person in the foreign country is telling you and work with it.
- You have to refer to both the international toll-free number in the foreign country as well as the domestic toll-free number it translates to.

Aside from these nuances, the rest of the procedure is essentially the same. The carrier needs accurate and precise call examples, as always, and you need to write down the trouble ticket number, take notes, and push the issue through by calling for updates every couple of hours.

Chapter 13

Troubleshooting Your Dedicated Circuits

In This Chapter

- Locating the level of the problem
- Identifying DS-3, DS-1, or DS-0 level variables
- Understanding circuit failure issues
- Addressing carrier trouble ticket questions
- Managing your dedicated trouble tickets
- Troubleshooting outbound dedicated issues, step by step
- Troubleshooting dedicated toll-free issues, step by step

Dedicated circuits save your business money when they are working, but if they fail, your company can come to a grinding halt. By identifying the problem accurately and by working with the information you uncover, you can greatly reduce your downtime.

Identifying the scope of a problem is always your first concern. When you know the level at which the issue exists, you can evaluate all the hardware and carriers that interact at that level and push forward with more tests. This chapter walks you through all the troubleshooting techniques, shortcuts, and pitfalls you might encounter with your dedicated circuits.

Keeping your objectivity is crucial, because the moment you close your mind to the possibility that the issue might be linked to a carrier or a piece of hardware, you begin reducing your chances of success. In the court of telecom, every variable is guilty until proven innocent. This philosophy should be the foundation of your cross-examination in the coming pages.

Identifying the Level of Your Problem

Dedicated circuits come in many sizes, and the first order of business is to isolate the level of the issue within the circuit. Here's a rundown of the basic levels of circuitry you might be dealing with:

- **OC-12 or OC-3:** Problems at these levels are usually related to large optical circuits. A problem at this level affects everything on the circuit. Alternatively, you may have a problem that just affects part of the circuit (one or two DS-1s on a specific DS-3), thus eliminating the OC-3 or the larger OC-12 (in which the OC-3 rides) as the possible source of the issue.
- **DS-3:** Problems at this level affect all the lower-level DS-1s and DS-0s. If red lights or alarms are flashing on your DS-3 multiplexer where the coax cable enters, and all of your DS-1s are down, you can trace the source to the DS-3 level.
- **DS-1 (also called T-1):** Problems at this level affect the entire DS-1 circuit and all the individual calls running on the circuit. If you have any individual DS-0 channel that is unaffected, your problem is not at the DS-1 level.

Don't get tripped up if someone uses the term T-1. T-1s and DS-1s are two names for exactly the same thing, like Bobby and Robert.

- **DS-0:** Problems at this level affect the individual call channel. These problems are either isolated to individual DS-0 level interfaces on your phone system, the DS-0 level hardware at your carrier, or somewhere within the public switched telephone network (PSTN). You know you're dealing with a DS-0 problem if the problem only affects specific calls; it won't affect the entire DS-1 circuit.

The good news is that troubleshooting anything above the level of a T-1 (DS-1) circuit involves the same carriers and quantity of hardware. The greatest change in variables occurs when troubleshooting issues at the T-1 level and individual channel (DS-0) level.

If your local carrier provides your dedicated circuit, be sure to include the carrier in your troubleshooting process. Most dedicated circuits for voice calls terminate at your long-distance carrier, and because of that, the local carrier has only a supporting role in the circuit. As you isolate dedicated circuits from the DS-3 or DS-1 level down to the individual DS-0 channel level, take note of the variables at work and who is responsible for each one.

Beware the misdiagnosed problem

Begin troubleshooting your dedicated circuit only after you confirm that your issue doesn't exist anywhere else. Use Chapters 11 and 12 to compare call types and confirm the source of your telecom problem. Your dedicated circuit only involves a few miles of cabling and a few pieces of hardware between your company and your carrier. If calls that don't use the dedicated circuit also experience the same problem, you need to report it to your carrier as a switched problem.

In fact, you may have a brand-new problem on your hands if you open a trouble ticket for your dedicated circuit and the problem could be effectively resolved as a switched issue. Put simply, troubleshooting the dedicated circuit places your circuit at unnecessary risk; you may end up taking down the circuit entirely, leaving your business without phone service for the duration of testing.

If your issue affects a variety of your long-distance calls, from switched outbound calls to dedicated inbound calls, I advise you (nay, I *entreat* you) to open your trouble ticket on the switched outbound issue, because it provides the greatest focus for your carrier.

Identifying circuit variables in circuits that are DS-3 or larger

Your multiplexer (MUX), the carrier's multiplexer (at the level of your circuit — DS-3 MUX for a DS-3 circuit, OC-12 MUX for an OC-12 circuit, and so on), and the physical connection between the two pieces of hardware handled by the provider of your local loop interact at the DS-3 level and higher.

If alarms ring on your DS-3 multiplexer, and the entire network span is down, a DS-1-level issue is not bringing down your circuit.

Similarly, if you have a problem with a T-1 circuit or on an individual channel on a single T-1, the problem can't be caused by the provider that supplies your local loop or with the DS-3 MUX. This is because DS-3 and DS-1 circuit problems don't interact at the individual DS-0 channel level. If your entire DS-3 has static on it, you must investigate the multiplexers on either end of your local loop. If you have a continuity issue and the span is down, the problem might lie with any of the cross-connects created by the local loop provider, or it could be a defect in the multiplexer at either end.

Identifying DS-1-level circuit variables

The primary elements of hardware interacting at the DS-1 level that can negatively affect the entire DS-1 circuit are

- Your DS-1-level multiplexer.
- The echo cancellers within the circuit.
- The card that provides the DS-level multiplexing at your carrier.
- If you have individual local loops for each of your T-1 circuits, your local loops are possible culprits.

Just as you do when troubleshooting at the DS-3 level, you must consider both the equipment that receives the circuit within your phone system and your carrier's hardware as the most likely culprits causing both quality and continuity issues. The copper wire that makes up the local loop is also a possible source of DS-1-level problems. If someone cuts through the cable that provides juice to your DS-1 circuit, everything riding through it is down hard. At the DS-1 level of a circuit, you encounter your first increase in software interaction.

If your local loop is a DS-3 circuit, or any circuit larger than the level of the issue you have identified, the source of the problem is most likely not the wiring or optical cable used to provide your local loop. If your DS-3 circuit has continuity, it is a safe bet that the DS-1s within your DS-3 are safely being received by your carrier and delivered to your phone system without incident. The local loop is essentially an empty pipe as far as your local loop provider is concerned; the local loop provider interacts only at the highest level of the circuit to maintain the clocking and keep the circuit stable. Your local loop provider ignores the T-1s or individual channels (DS-0s) within the local loop, except for the possible interaction of echo cancellers at the T-1 or individual DS-0 level if the circuit is an ISDN circuit.

Identifying DS-0 or individual channel issues

Individual channel issues generally indicate a problem with either your multiplexing hardware or the cards your carrier uses in its switch to perform the multiplexing. You can identify a DS-0 issue pretty easily. If your DS-1 is working fine and many channels on the DS-1 can send and receive calls without any problem, there are really only two main variables and one minor variable to evaluate in determining the source of the problem.

The main variables at the DS-0 level are the multiplexers on either side of the circuit that break down the T-1 into individual channels. If your third DS-0 on your T-1 has static, is unavailable, or if you can't seize a dial tone on it, either your hardware or your carrier's hardware is failing for that channel. If you can isolate both pieces of hardware and the problem doesn't clear up, the only other likely candidate is possibly an *echo canceller* somewhere in your circuit that is malfunctioning or dying a slow death. The purpose of an echo canceller (also referred to as an *echo can*) is to eliminate the echo heard on a call. If the device is failing or was provisioned incorrectly, it can prevent your call from completing with good quality, or at all.

You can also encounter problems at the individual call level caused by software compatibility in the outpulse signaling and start. If your system is configured for loopstart and your carrier is set for E&M Immediate, you will have problems making calls. This type of configuration issue causes calls to suddenly disconnect, fail to connect at all, and otherwise fouls things up. Configuration issues at the individual call level can also cause frame slip errors and errors that can take down your entire circuit. Chapter 8 covers line coding and framing in detail, as well as the symptoms you will encounter due to frame slips and errors on your circuit.

Categorizing the Nature of Your Problem

Dedicated circuits can be intimidating because you are responsible for half of the hardware that makes them work. On the other hand, the majority of problems you will likely encounter originate in your long-distance carrier's switched network, far beyond the end of your dedicated circuit.

The same types of call quality and completion problems that you encounter on switched calls can also affect the calls that pass through your dedicated circuit. The joy of dedicated circuits is that you can test for either problem in the exact same way: by isolating and eliminating each variable as you go through your circuit.

There might be other issues that are associated with your dedicated circuit, such as dropped calls or intermittent echoes, that you troubleshoot based on multiple call examples, but overall, the steps for resolution on dedicated circuits are the same. Chapter 11 covers the requirements for call examples and why they are necessary in troubleshooting.

Understanding dedicated call quality issues

If you have call quality issues that exist only on your dedicated circuit, you can track down their sources much faster than you can call quality issues that you notice in your carrier's switched network. The calls running through your carrier's switched network may have 20 or 30 dedicated spans and pieces of hardware associated with them, each one requiring tests and validation. If the quality issue is truly confined to your dedicated circuit, on the other hand, your problem is confined to a finite space.

Quality issues such as static, low volume, *clipping* (where you hear continuous split-second gaps in your conversation), and possibly even echo, leave a trail of clues and evidence for you to follow as you troubleshoot. All these issues can be the result of poorly provisioned hardware; maybe one section of the circuit is set up for loopstart signaling and the rest of the circuit is set up for E&M Wink signaling (read more about these settings in Chapter 8). A piece of hardware wearing out and failing could also cause quality issues. Confirming what's what with defective hardware is not difficult because all network switches in North America have computer files that can be activated to collect performance data on the circuit.

Files saved by computer software to monitor call performance are unoriginally called *performance monitors* (PMs), and they collect information on the quantity and type of errors on a dedicated circuit. Your phone system may have a similar feature, allowing you to collect the errors experienced by your hardware. If your hardware isn't as advanced, you can request the information from your carrier by calling to open a trouble ticket. PM software categorizes errors in several categories. If there was no error on a circuit in a particular category, the "0" error count is listed. The quantity of each type of error guides you to other areas of inquiry when you see them. Here are some common PM errors:

- **Erred seconds:** You find *erred seconds* on a circuit with a minor problem. An erred second indicates that for the duration of one second, the communication of the circuit was distorted. The overhead information may have been lost for that specific second because some piece of hardware experienced a low-level, intermittent electrical short. You might experience minor static or quality issues that are more of an annoyance than a concern.
- **Severely erred seconds:** You can find severely erred seconds on a circuit with large issues. The severely erred seconds indicate a larger issue whereby information was not only missing, but replaced with aberrant data. These errors often cause static or line noise so severe that you can't hear the person you are calling. If your circuit has severely erred seconds but you aren't experiencing any regular quality issues, you might instead notice that your calls disconnect prematurely or that your entire circuit drops unexpectedly.
- **Framing slip:** Framing slips signal a configuration problem, generally a timing issue where your hardware is attempting to correct a lag behind

the master clock of your carrier. If you aren't *clocking* off your carrier (see Chapter 8 if the term clocking isn't clicking for you), or if your hardware is set up for an outpulse signal and start that doesn't match your carrier, you can expect frame slips.

The bad news about frame slips is that they may be small, minor, and go unnoticed by you as they accumulate, until the point where they drop your circuit.

- **Unavailable seconds:** Unavailable seconds are just that, seconds of time when the network thinks your circuit is unavailable. Unplugging your hardware generates unavailable seconds in your carrier's network as easily as a fiber cut or an accidental removal of your cross-connect by your local loop provider.

One of the simplest things you can do to see whether a problem on your dedicated circuit can resolve itself is to reboot the whole system. Sometimes all you need is PFM *(pure friggin' magic)* to get everything to right itself. If you decide to reboot your system, make sure that your carrier pulls the PMs before you reboot your hardware; this way, you can prevent confusion. Even if you have a clean circuit, as soon as you power down your channel bank and reboot it, you create a handful of errors on the circuit. Even if you have a legitimate issue on your circuit, rebooting has the potential to mask the true issue and enlarge the number of errors your carrier finds. For more information about PFM, check out Chapter 16.

Without controlled testing, your carrier won't know which piece of hardware is generating the errors, or when the errors were created. Errors that exist in the PM files don't have time stamps next to them, only total error counts for a given period of time. If the PMs show your circuit received 176 erred seconds and 56 severely erred seconds in the past 24 hours, you have no way of identifying whether all the errors occurred in one group as your circuit took a hit, or if they were generated periodically over the past 24 hours. The only way to validate the frequency of errors is to purge the PM files and check it every 15 minutes, in order to determine the rate at which the errors are generated.

Understanding circuit failure issues

The larger the problem, the easier it is to find. If your circuit is completely down and alarms are sounding on your hardware, the bad news is that you may have a large problem, but the good news is that finding its source is pretty easy.

Most dedicated circuits are only about 4 or 5 miles long and have a handful of hardware interfaces in them from end to end. If any single piece of hardware or cabling that handles your circuit fails, your circuit will fail along with it. You know your circuit is in failure when

- **Your multiplexer and/or CSU have active alarms.** This can include flashing red lights where there used to be solid green lights, and/or an annoying beeping sound.
- **You can't make calls or receive toll-free calls on your circuit.** On the other hand, if outbound calls on your circuit are affected, but you can receive dedicated toll-free calls, the problem is most likely with your phone system, not the circuit.
- **Everyone in the office is yelling at you to fix the phones.** A heightened level of telecom awareness and irritation in your office is a strong indication that your dedicated circuit isn't working right. In this case, just tell everyone not to worry and show them your *Telecom For Dummies* book. When they see it, they will be instantly calmed and realize that you are in complete control of the situation.

Even after you find the source of a problem, repairing it can take many more hours. Dispatching a technician from your local loop provider requires coordination to have someone at your office let the technician into the building. Depending on the technician's schedule and yours, the logistics of setting up a dispatch might push the repair out to the next business day. To prevent your problem from dragging on, the second you realize that you have a system failure (whether it occasionally bounces back or not), push to get it into the queue for a technician as quickly as you can.

Got a headache? Then you must have a bouncing circuit

Circuits don't always fail and stay down. Sometimes a circuit may *bounce,* meaning that it fails and then comes back to life, only to drop again and repeat the cycle a few minutes later. A *bouncing circuit* is problematic, because it requires you to make a decision: You must either limp along and live with the circuit failing every so often or you must let your carrier take the circuit down for 30 to 60 minutes to test the circuit and find the problem. You hardly have any choice at all if the circuit drops so frequently that it might as well be permanently down. On the other hand, if you have just a few hours to go before you can legitimately close down business for the day, you might be better off hanging in there and suffering through the rest of a very long afternoon. Then after everyone goes home, release the circuit for testing.

Opening a Trouble Ticket for Your Dedicated Circuit

After you have some preliminary information about the trouble with your circuit, you need to call your carrier and open a trouble ticket. It's at this time that you will realize the benefits of having a technical cut sheet. If you have not made one yet, now is a great time to do so. Check out the Cheat Sheet at the front of this book to set up a cut sheet today.

Going through the basics

If you don't have a cut sheet, you need to write down the circuit ID for every circuit that is having problems, at a very minimum. Your carrier needs this information to identify your circuit and begin the troubleshooting process. In addition to your circuit ID, your carrier might also ask you the following questions:

- **Does your hardware show any alarms?** If so, what color are they and where are you seeing these alarms (CSU or multiplexer)? Both the alarms and their location indicate to your carrier what kind of problems your circuit is experiencing. If you have red alarms on your multiplexer, the carrier knows that the circuit is down hard and that your business has no phone service at all. On the other hand, if your CSU is experiencing yellow alarms, your circuit could be intermittently down and bouncing back. If you don't know this information, you can simply tell your carrier that you haven't checked yet, but that you still want to open the trouble ticket.
- **Have you rebooted your multiplexer/CSU, and if so, how did it affect the problem?** You might temporarily resolve some problems by rebooting your hardware. Rebooting enables your hardware to refresh the connection to your carrier and synch up. The only variable you are testing when you reboot is the multiplexer. If the problem clears when you reboot, your hardware is probably the source of your issue. If it persists without any change when you reboot, the source of the problem could be within your carrier.
- **What are your hours of operation?** In the event that a technician has to be dispatched to your office to fix the problem, your carrier needs to know when someone will be there to let in the technician. If you have an after-hours employee who will be on-site, be sure to provide that employee's name and phone number (if the system is down hard, provide a cellphone number). You don't want the technician to arrive and

have no idea where to go. If the receptionist or building security has no idea you have a technician on the way, but the technician says, "I was told to ask for Janice Jackson," he or she is more likely to get through the front door.

- **When did you first notice the problem?** This information gives your carrier direction in the search. If there was a large outage that hit your area an hour ago, and that is roughly when you noticed the situation, the carrier can easily combine your issue with the overarching trouble ticket (sometimes called a *master trouble ticket*). Of course, your problem may not be related to a larger issue, but offering a time frame does give your carrier some indication of events that occurred that may have caused your issue.
- **Who is the site contact?** Your carrier needs to know who to call for updates. Please provide a direct phone number, a cellphone number, and a secondary means of contact if your contact plans on being unavailable at any time during the day. It's painful to pick up a voicemail from your cellphone after you have lost four hours because you ran into a meeting, got busy, and didn't have time to listen to it until the end of the day.
- **Is the circuit released for intrusive testing?** *Intrusive testing* is the most direct way your carrier can investigate an issue with a dedicated circuit. The main downside of intrusive testing is that your circuit will be completely down while your carrier takes over every channel. Of course, if the system is already down, you have nothing to lose. On the other hand, if your circuit is active when you release it for intrusive testing, any active calls will be disconnected when the carrier initiates the test. See the section, later in this chapter, "Step 2: Intrusively testing: Looping the CSU" for more information about timing your test.

Letting your channels be your guide

When you open a trouble ticket, it's always helpful to ask the carrier for a *circuit snapshot.* A snapshot of the circuit is the disposition of each channel on your circuit. Even if you think you have a DS-1 level issue, it's always a good practice to ask for this information. You might believe that your entire circuit is being affected, but a snapshot could reveal that only half of the channels are impaired. With basic facts from the PMs and a circuit snapshot, you can begin setting realistic expectations for resolution of your service. Additionally, you can isolate individual DS-0 issues to the specific channels experiencing the problem. To solve your problem, you might be able to reboot your phone system to bring the channels back into service, or have your phone system deselect them as available lines. The key is using every available bit of information your circuit provides. Taking in the disposition of your individual DS-0 channels on your DS-1 circuit is a quick and easy way to glean what is happening to your circuit.

Remembering the first rule of troubleshooting

The first rule of troubleshooting is to keep an open mind. Suspecting that a problem can be attributed to a particular source is fine, but don't be so sure you're right that you lose perspective. All the circuit states I describe in the following sections do suggest a likely source for the problem you're experiencing with your dedicated circuit. Most of the time, for example, a channel in RMB (remote made busy) state is having hardware troubles. However, there is that tiny chance that a glitch within your carrier is causing it to send the channel a message, or triggering the channel to sit in RMB. Alternative scenarios are always possible, not just for channels in an RMB state, but for all the circuit states listed in this chapter. Use the information to guide your troubleshooting, but don't be so hung up on a single right answer that you discount the possibility that something else is causing the issue.

Idle: IDL

Your channels are in an *idle* state when they have access to your carrier's network, but don't have an active call. This is the state you expect your channels to be in when everything is working fine and the circuit is waiting to process a call.

The circuit must be considered idle by both your carrier's technology and your own phone system in order for a call to come through on the line. If, according to your carrier, the channels are idle, but your hardware is unable to seize the channels for some reason, one end of the circuit isn't speaking to the other. Open a trouble ticket. The problem might be between your phone system and your multiplexer, or your multiplexer and your CSU. If your carrier sees the circuit as active and you can't seize a channel, there must be a piece of hardware interacting with your carrier to make the carrier think everything is fine.

Call processing busy: CPB

Call processing busy or *CPB* is the healthy state of a channel with an active call on it. Under normal circumstances, your active circuit has several channels in the CPB state, indicating active calls on the specific DS-0s. The rest of the channels are idle, waiting to accept a call. A circuit with channels in CPB is obviously connected to the carrier, because only a call that is connected to the network can establish the channel in this state.

Remote made busy: RMB

Remote made busy or *RMB* is a common disposition for a channel if you're dealing with hardware issues. The carrier may see a DS-0 in this state when your hardware independently locks out the channel. Channels in an RMB state can sometimes be corrected if your carrier takes down the specific channel and resets it (this process is commonly called *bouncing a channel*). If your carrier can't bring the DS-0 back into service by bouncing it, the only way to restore the channel is for you to reboot the hardware. If this condition is chronic, contact your hardware vendor to test the hardware. It's better to find the problem when it's minor, than to let it go until the point at which the channel won't come back to life after you reboot.

D channel made busy: DMB

The *D channel made busy* state is used only with ISDN circuits. Only ISDN circuits have D channels designated for signaling, and that is why non-ISDN circuits can't have channels in this state. When the D channel made busy state appears on an ISDN circuit, the entire circuit is down. The D channel handles all the logistics of call setup and teardown on an ISDN circuit, and without the overhead, the circuit can't function.

This condition can be the result of your hardware having issues (see the section on RMB, earlier in this chapter), your carrier having issues, or an ISDN protocol mismatch.

Installation made busy: IMB

Installation made busy, or IMB, is a protected state that is usually imposed by your carrier on an entire circuit to prevent it from generating alarms in the carrier's network. For example, if you unplug your hardware from the circuit, your carrier immediately sees alarms at its end of the span. The technician who is listening to the piercing screech of the alarm doesn't have your contact name and phone number, so he or she does the easiest thing to end the maddening alarm. Usually, the easiest thing to do is to *busy out the circuit,* or place the circuit into the IMB state. The bad news is that when you plug your hardware back in, you won't be able to use it because the carrier has been busied it out.

Every carrier has its own procedure for placing a circuit into the IMB state. The process might be manual, with a policy to wait until the circuit has been in alarm mode for eight or ten hours before busying out the circuit. Sometimes, however, the process is automated; you may have as little as an hour before the carrier automatically busies out your system.

Carrier failure: CFL

Carrier failure is like IMB, because it generally affects your entire circuit, and not just a few channels. The bad news about a circuit in carrier failure is that this state commonly indicates a substantial problem that needs to be addressed pronto by your carrier. If your CSU fails, or if someone accidentally cuts through the wiring of your circuit with a backhoe, or if your carrier's switch sustains a direct lightning strike, your circuit is in carrier failure.

Avoiding permanent IMB status

To prevent your circuit from spending more time in IMB than it needs to, open an *information-only trouble ticket* with your carrier before unplugging your hardware for any reason. This type of trouble ticket alerts the carrier to the fact that you will be removing your hardware from the circuit and gives you a trouble ticket number to reference when you call into have the circuit taken out of IMB.

If you don't want to go through this effort, simply place a loopback plug in the T-1 jack of the circuit from which you're disconnecting. This simple action tricks your carrier into thinking there is hardware connected on your end of the circuit for as long as the loopback plug is in place. This isn't an option if you have ISDN circuits, because your carrier needs to chat with your D channels to ensure that the hardware is active. The ISDN protocol is generally intelligent enough to know that it is speaking to itself and eventually places the circuit in alarm.

Removing a circuit from IMB takes about five minutes, after you reach a technician with the required credentials to make the change. The greatest potential delay comes from wading through the customer service department and waiting for the magic technician with the access to the switch to call you back.

Carrier failure indicates a more serious issue and generally takes several hours to diagnose. After the technician finds the source of the problem, it can be another 4 to 24 hours before your circuit is back in service. Keep this in mind when you set expectations. (In other words, it's time to have a chat with the head of your telemarketing department.)

Managing Your Dedicated Trouble Ticket

After you have opened your trouble ticket with your carrier, be sure to write down all the following information to make it easier to manage the issue:

- The trouble ticket number.
- The name and contact information of the person at your carrier who opened the ticket.
- The time and date the ticket was opened.
- The disposition of the DS-0s of your circuit (if available).
- The errors listed on the performance monitors, if any (if the customer service representative can access the file).
- Any testing you completed and the results of the tests. (Did your carrier bounce the circuit by taking the circuit out of service and restoring it? Did that solve the problem?)

The final bit of information your carrier should provide to you is when you can expect a call back on your ticket, and who will be calling. Regardless of the time frame you're given, write it down, and schedule to call the carrier back at that time. If the technician calls you before the deadline, great, but if he or she doesn't, you need to call back in order to escalate your issue.

Getting the Basics of Dedicated Outbound Troubleshooting

When you troubleshoot a dedicated circuit, you do it in the reverse order that you troubleshoot switched phone lines. If you have a problem calling Hoboken, New Jersey, on a switched phone line, for example, you should perform your troubleshooting before you call your carrier and open up a trouble ticket. When it comes to dedicated circuits, however, you begin your troubleshooting after you open the ticket.

The reason you should open your trouble ticket forthwith is because you don't have many testing options on a circuit without the aid of a technician from your carrier. You might know that your circuit isn't working, but only your carrier can tell that 10 of your channels are in an RMB state and 14 are in IDL. Even if you know the problem you are experiencing is caused by your hardware, you should still open a trouble ticket to receive help validating the source.

The troubleshooting steps in this chapter deal with a simple dedicated long-distance voice circuit with one local loop provider. If you have a type 2 or type 3 circuit that involves more than one carrier to deliver the local loop, the troubleshooting process is more involved and beyond the scope of this book. One thing I do suggest is that you put all the carriers involved in your circuit on a conference call at the same time.

Step 1: Rebooting your hardware

If your circuit is down hard, either in an RMB or CFL state, it's always good to begin your troubleshooting by rebooting your multiplexer and/or CSU.

The two pieces of hardware that are actually talking to each other on the circuit are your multiplexer and the computer card at your carrier that functions like a multiplexer, so rebooting the hardware on your side may resolve the issue. If rebooting brings your circuit back up, you should still follow up with the trouble ticket to determine the root of the problem. Most problems don't happen once and then disappear forever. If troubleshooting indicates a hardware problem, you ought to replace the faulty hardware while it's still limping along, instead of waiting for it to fail completely.

Understanding your trouble ticket options

You can open special trouble tickets for specific issues. Not only can you open a ticket for repair, you an also open an *information-only ticket*, or a *tech assist ticket*. Opening an information-only ticket allows you to officially relay information to your carrier. If you want to tell your carrier that you will be disconnecting your hardware for the weekend because you are moving your phone closet to another suite, for example, the informational trouble ticket relays the information and gives you quicker access to a technician if you need one.

Use a tech assist ticket to schedule nonemergency testing between your carrier and your hardware vendor. If you are replacing your hardware, for instance, you should coordinate with your carrier to be either on the phone with your hardware vendor while making the change, or at least to be immediately available to the vendor.

The last thing you need when you encounter a problem during such a change is to wait hours to speak to a technician. You might not be inconvenienced if the maintenance takes place after normal business hours, but your wallet will still be skinnier after you have to pay for your vendor to sit around and wait for the call.

Rebooting your hardware isn't helpful for troubleshooting call quality issues, because they will generally be the result of something being set up wrong or a progressive failure in a section of your hardware. The first problem could be anything from a protocol mismatch to a configuration mismatch.

Here are some important rebooting do's and don'ts:

- **Do make sure more than one person knows how to reboot the phone hardware.** You and someone else in your office (your backup) should know how to reboot every piece of your phone system. It's a waste of time and money to call out your hardware vendor to push one button. If you don't know how to reboot your hardware, ask your hardware vendor to show you. If it's a complicated procedure, have the vendor write it in black pen on the wall of the phone room next to your hardware. It's well worth having graffiti on your wall to be able to confidently push a button without worrying what will happen (besides, graffiti art is cool).

- **Don't be in the dark about whether your phone system will restore to default settings.** It's uncommon, but not unheard of, for hardware to restore to default settings after being rebooted. When you discuss recycling the power of your PBX, multiplexer, and CSU (as well as any other servers or devices attached to these three pieces of hardware) with your vendor, be sure to ask about possible side effects of rebooting. While you are at it, also ask about fixing these side effects if they come up. The solution might be very simple, or it might be, "don't touch anything; just call me." Either way, this information is good to know.

- **Do make note of the frequency of your reboots.** If your circuit drops on a regular basis, be it once a month or once every three months, but immediately comes back to life when you reboot your hardware, your hardware might have a timing issue. A timing issue can be caused by your hardware utilizing a clock for the point of reference on your TDM circuit that doesn't come from your carrier, or the section of your hardware that handles the clocking on the circuit may be failing. In either case, timing issues cause the phone system to accumulate small slips within the circuit that eventually cause the circuit to fail. You might ignore the problem because your circuit comes back to life as soon as you reboot, but the issue isn't solved. It will progress over time, and the time it takes for your circuit to fail will gradually become shorter. If this scenario sounds familiar, contact your hardware vendor to check your phone equipment. If this doesn't make a lot of sense to you, read up on the clock source requirements of dedicated circuits in Chapter 8.

Step 2: Intrusively testing: Looping the CSU

Your carrier can *intrusively test a circuit,* which renders the circuit useless to you during the duration of the test. Every channel of the circuit being tested must be engaged so you won't have any room left to either make or receive calls.

As soon as your carrier begins intrusive testing on your circuit, all the active calls on your circuit will be disconnected. The testing environment usurps all the bandwidth. Your hardware can't grab any DS-0 to make or receive a call during an intrusive test. If your circuit is already down, you should release it for intrusive testing. If the circuit is bouncing, or only slightly impaired, you need to make a judgment call. Can your business stand to have the circuit down for 30 to 60 minutes now, or should you wait until after your office closes to release it for intrusive testing?

You don't have to release your circuit for intrusive testing at your carrier's discretion when you open your trouble ticket. If your circuit is bouncing, ask the carrier to do some preliminary work first (such as accessing the system's performance monitors). You can always release the circuit for testing after hours, say at 7 p.m. central standard time. This way, you can get basic information, and when you get in tomorrow morning, you can get more information about what the carrier found during the intrusive test.

An intrusive test involves looping the CSU. *Looping* is a general term used in telecom that describes the act of sending an electrical signal to a specific piece of hardware where it's bounced back to your carrier. Your carrier technicians actually send a stream of data down the circuit, possibly consisting of *all 1s* or *all 0s.* If every "1" the technician sends in a test effectively hits the desired endpoint and comes back to your carrier, your circuit has solid continuity from your carrier to that point.

If the idea of looping doesn't make sense to you, imagine it another way. If you have a garden hose with leaks, the easiest way to find them is by placing a spray nozzle on the end of the hose and closing it so no water can escape. After the end is sealed off, you can check for leaks by turning on the faucet and looking for any water springing out of it. Intrusive testing does the same thing, but with data. If you send a series of 27 consecutive 1s down the circuit and only 25 return, you know that a pair of your 1s were lost somewhere between the point at which the data originated and the looped piece of hardware that is sending it back to you. Losing data means that you have a problem in that section of your circuit.

There are many tests that your carrier can use when looping a piece of hardware. They have names like *QUAZI, 4 and 8, 6 and 2,* and my personal favorite, *all 1s.* Some tests are more rigorous on a circuit than others, so have your carrier try a few of them if you don't feel confident with the results of the first test.

Intrusively testing a dedicated circuit follows a very methodical sequence. Your carrier's first test is attempting to loop your CSU. If you don't know what a CSU is, please check out Chapter 4 for info on your hardware requirements for a dedicated circuit. If you are more concerned with how the CSU is positioned in the circuit, check out Figure 13-1 to see where it lives in the dedicated circuit landscape.

Dealing with synch issues caused by Dialogic cards

If you are using Dialogic cards as the multiplexing interface on your side of the circuit, you may have problems when you reboot your hardware. There is a known issue that exists between Dialogic cards and some switches that the carriers use. If you find that your Dialogic cards don't automatically synch up when you reboot, or if individual channels lock up when you dial to a bad phone number (where you receive a recording or a fast busy signal), you need to enable the *CRC-6 setting* on your Dialogic card to resolve the issue. You might also need to reboot your Dialogic card before the programming change takes effect. After the setting is enabled, your channels shouldn't fall out of sync with your carrier any more.

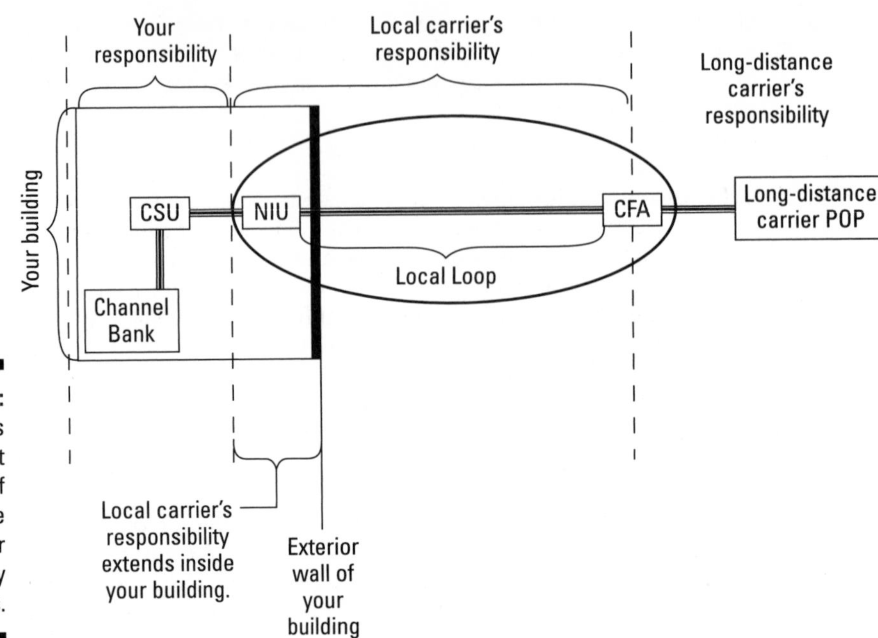

Figure 13-1: The CSU is the first piece of hardware your carrier intrusively tests.

Figure 13-1 shows that the CSU is the first piece of hardware your carrier identifies as your responsibility. There is a cable from the NIU to your CSU that is also your responsibility, but it isn't generally loopable, so the CSU is the default piece of hardware used as the first point for testing.

Not all CSUs are identical. Your CSU might not be a free-standing piece of hardware as portrayed in Figure 13-1, and might instead be an integrated card housed within your multiplexer. It's uncommon, but not unheard of, for a CSU to not be loopable. This isn't a huge stumbling block, but you simply need to be aware of this bit of trivia. Your carrier might try to imply that your hardware is defective, because the carrier can't loop your CSU.

If looping the CSU fails

Here are the possible meanings of a failure to loop the CSU:

- If you can't loop the CSU, this is a strong indication that there are problems somewhere in your local loop.
- If you are testing to identify the source of a circuit failure issue, and your carrier can't see the CSU to loop it, there's a break in the circuit somewhere in the area being tested.

- If you are troubleshooting for a call quality issue and you can loop the CSU (indicating that you have electrical continuity) but you receive errors, some piece of hardware within the testing area is probably dying a slow death.

These tests don't indicate that your CSU is the problem; they simply indicate that something between your CSU and the CFA point is the problem. The only way to isolate the issue is to proceed to "Step 3: Looping the NIU."

If looping the CSU is successful

If your carrier can loop your CSU without a problem (for circuit failure issues) and can run test patterns to it without taking any errors (for circuit quality issues), the entire section of the circuit you are testing is clean. Looping your CSU generally causes the equipment on the other side of it, usually your multiplexer, to idle up all the channels.

Looping the CSU not only bounces the circuit signal back to your carrier, but also bounces the signal that it's sending back to your multiplexer. If your MUX doesn't idle up to the back of the looped CSU, your next course of action is to have your hardware vendor test your MUX to determine whether it's the point of failure.

Looping a piece of hardware doesn't confirm that everything is wired correctly to the device being tested; it simply confirms that you have electrical continuity. See the following section to make sure your wires aren't crossed.

Making sure your crossover cables aren't straight-through cables

Dedicated circuits have designated wires to transmit and receive signals. Even if these *transmit* and *receive* wires are crossed between the NIU and the CSU, you can still loop the CSU; but in that case, the T-1 level of your circuit won't idle up. If this situation occurs, you need to check the configuration of the cable you are using to connect your NIU to your CSU.

There are two main varieties of cables in telecom, *straight-through cables,* and *crossover cables.* The wires that make up cables are numbered for easy reference, allowing you to understand why they have their specific label. Figure 13-2 shows a four-wire circuit, with each wire identified by a number. You can see how the black wire #1 on the straight-through cable remains wire #1 at the far end, whereas it ends up as wire #4 on the far end of the crossover cable. Needless to say, it's very easy to wire a cable incorrectly and end up using the wrong wires for transmit and receive.

Straight-Through Cable

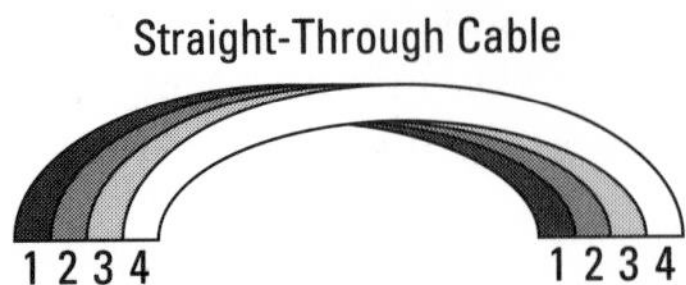

Crossover Cable

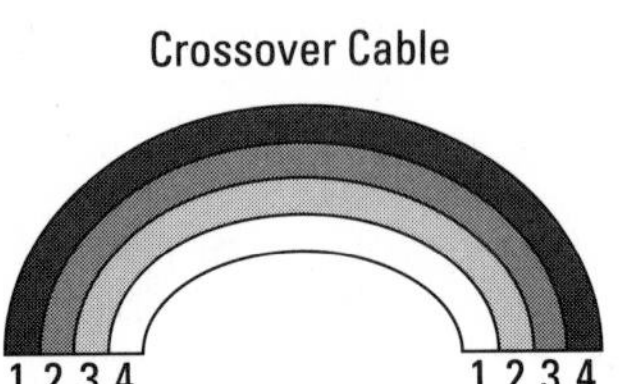

Figure 13-2: A comparison of straight-through and crossover cables.

If the CSU can be looped and you have checked the cabling, but you still don't have dial tone, you need to confirm the configuration of your circuit. In spite of the fact that you may have confirmed the line coding and framing is set for B8ZS/ESF on both ends of the loop, that still doesn't prove that a piece of hardware isn't programmed incorrectly. Your carrier can validate that the line is B8ZS/ESF by sending *all 0s* down the circuit. If the zeros come back without errors, you can rest assured that all the hardware it's hitting is configured for B8ZS/ESF. Your carrier can run the same type of test for AMI/SF by sending down *all 1s*. When you've validated the line coding, you should confirm the outpulse signal and start (E&M Wink or Immediate, loopstart, or groundstart) on both your hardware and the carrier. If you want more information about dedicated circuit line configuration options, like B8ZS/ESF, see Chapter 8.

If the CSU can be looped, and the configuration has been confirmed, the next variable to examine is the hardware on both ends of the circuit. Just as you have a multiplexer that breaks down your T-1 into 24 individual channels, your carrier has a card in its phone switch that does the same thing.

It's very unlikely that a T-1 level card in your carrier's switch will fail, but it does happen. If you have more than one circuit and only one of the circuits has been acting up, ask your carrier to reassign one of the known bad circuits to the card that is working on the good circuit. If your hardware instantly comes to life when you are on the known good card, the other card is obviously bad (that was the only variable you changed). If you make the change and your circuit is still down, your hardware is probably the source of the trouble. Move on to the following section.

Using a T-1 test set

If all the previous tests don't reveal the source of your problem, your next step is to replace your hardware and see whether that brings the circuit up. If this doesn't resolve your issue, you need to call out a hardware technician with a portable *T-1 test set,* possibly a Phoenix test set model 5575 or a T-Berd test set made by TTC.

If you have a complex phone system, or if your business is in any way based on providing phone service, whoever is servicing your phone system should have a piece of testing hardware. Test sets look like a cross between a toolbox and a 1930s lunch box. They range in price from a few hundred dollars for a used Phoenix 5575A, to several thousand dollars for a fully fleshed out TTC T-Berd or TTC FireBerd with all the available ISDN options.

Having a test set for your dedicated circuit is very helpful, so chat with your hardware provider and see what is the best test set for your application and budget. A refurbished Phoenix 5575A might be all you will need and is well worth the $200 you might end up spending for it on eBay.

If you don't want to buy your own test set, you can always have a hardware technician come out and do the work for you. Call your hardware vendor and set up an appointment as soon as possible.

With a test set, you can run the continuity tests that your carrier typically performs. This allows you to begin testing as soon as you have a problem on your circuit, without waiting for a technician to call you back. After you have a technician on the line testing, you can validate his or her findings, and provide the technician with much greater information on your circuit. You can plug the test set into your circuit between the NIU and your CSU for the greatest vantage point. By placing the test set in your circuit at this location, you bypass your entire phone system. This is a tremendous benefit because the test set acts as an insulator between your hardware and your carrier. If your test set is monitoring the signal coming in from your carrier and it shows that the carrier is sending B8ZS/ESF signaling rather than AMI/SF, there is no way the problem can be related to your multiplexer. All the hardware behind your test set is hidden by it and invisible to your carrier. The minor investment will pay off, because it will no longer be possible for your hardware vendor and your carrier to point fingers at each other. The test set is an independent piece of hardware that you can use to replace your multiplexer in the circuit, and can allow you to dial out on an individual channel if your test set has the correct options. If your problem clears when your carrier is connected to your test set rather than to your multiplexer, your hardware is at fault and your next course of action is a detailed check of your multiplexer and the network behind it.

Step 3: Looping the NIU

If your carrier can't loop your CSU, the next step to attempt is to drop back one step closer to the CFA point of your circuit and loop your NIU (or one step closer to Nirvana, depending on your perspective). The NIU is the last piece of hardware that your local carrier installed, as shown in Figure 13-3.

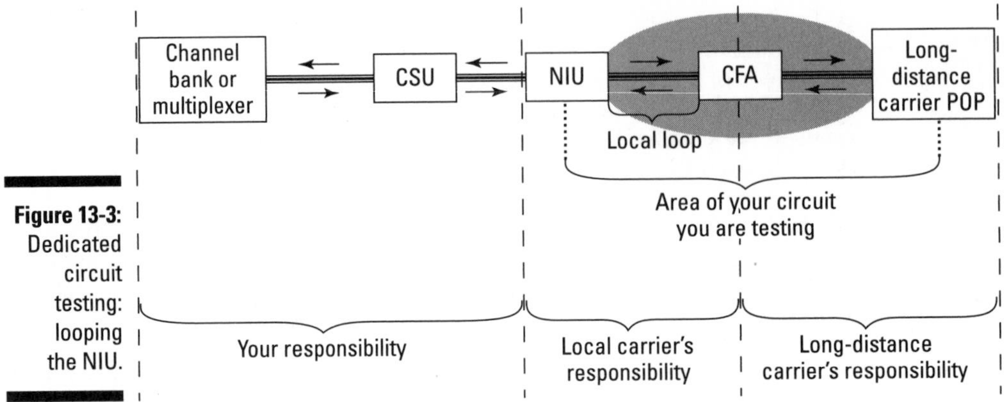

Figure 13-3: Dedicated circuit testing: looping the NIU.

By looping the NIU, your carrier can test to the back end of the NIU where your local carrier wires into it. Your goal when methodically validating smaller sections of the circuit from the CSU back to your carrier is to find the section that is without errors or problems. Stepping back your continuity tests (for circuit failure issues) or error tests for a clean span (for quality issues) to a section without issues gives you a starting point on the circuit that you know is good. After you find that good section, you can focus your efforts on the next variable, working your way closer to your CSU, which is the hardware or the span of the circuit that's most likely the source of the issue.

If you can loop the NIU

If your carrier can loop the NIU but not the CSU, the problem lies somewhere between the back end of the NIU and the front end of the CSU. The cabling or the CSU has a break in the wiring (if your carrier is seeing a lack of continuity on the line, often referred to as *an open*) or is shorting out (if the carrier sees errors).

If you test the CSU and it can be looped, you need to ask whomever provided the inside wiring to repair the trouble. If your local carrier pulled the inside wiring, it might be quicker to have your hardware vendor validate or rerun the cabling, instead of waiting the one to five days to have your local carrier dispatch someone to your site. To test the inside wiring, proceed to "Step 4: Looping to your T-1 jack."

Getting the scoop on loops

There are two varieties of testing loops in the world of telecom: *soft loops* and *hard loops.* CSUs and NIUs generate soft loops, which use software to bounce the signaling on a circuit back to its point of origin. You create hard loops by physically wiring together the transmit and receive lines on a circuit. You typically place a hard loop in a circuit by using a loopback plug, but both loops perform the same function.

If you can't loop the NIU

If you can't loop the NIU, there is a problem in the local loop, or possibly within your long-distance carrier's network. The only way to isolate the issue is to continue working back on the circuit towards the CFA point until you find a good section that doesn't have errors. In this case, proceed straight to "Step 5: Looping the CFA point" for additional troubleshooting.

Step 4: Looping to your T-1 jack

It's unlikely that your CSU is connected directly into your NIU. These two pieces of hardware might be in different sections of your building, connected by the inside wiring that ends next to your phone system in a small T-1 phone jack (called an RJ-45 jack). Figure 13-4 shows how the RJ-45 jack sits between your NIU and your CSU at the end of your inside wiring.

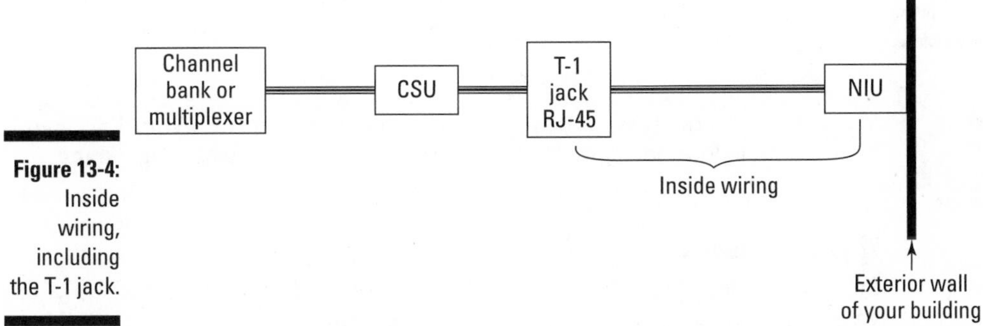

Figure 13-4: Inside wiring, including the T-1 jack.

In order to test to the RJ-45 jack for continuity and errors, you have to plug into it with a test set, or use a *loopback plug,* which is a much cheaper alternative. Loopback plugs are available in both male and female versions so that you can test your cables and the T-1 jack on the wall.

The appendix at the end of this book offers detailed descriptions of the loopback plugs. They are well worth the 50 cents of hardware it takes to make them (although the crimper tool used to clamp down the pins may cost up to $50).

When you place the loopback plug in the jack, you allow your carrier to continue to send signals through the NIU to the phone jack. If the carrier can see the loopback plug and run test patterns without losing any data or receiving errors, you know that the section of the cabling to that point is solid. If your carrier can't see the loopback plug, or is taking errors to it, but can loop the NIU without a problem, your trouble lives in the wiring between the NIU and that T-1 jack. Figure 13-5 illustrates the portion of the circuit you have to test next.

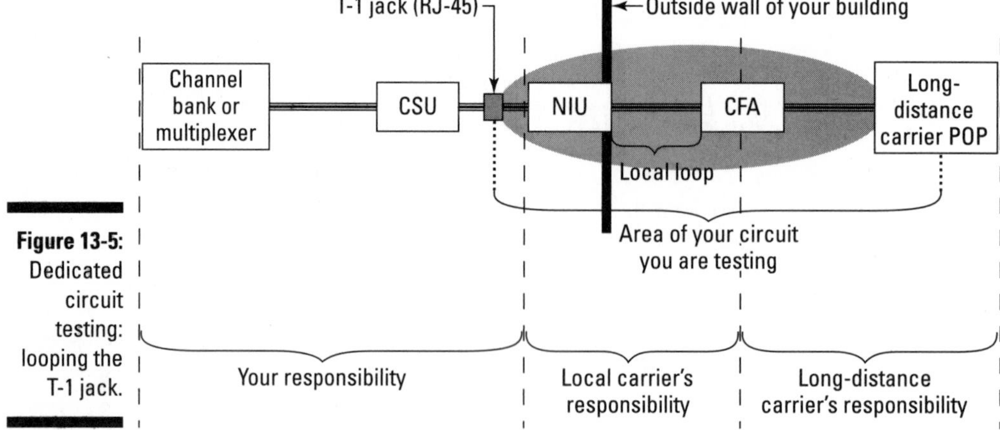

Figure 13-5: Dedicated circuit testing: looping the T-1 jack.

You can extend the test to the end of the cable that feeds into your CSU if you have a female loopback plug to receive the T-1 cable that was plugged into your CSU.

If you can loop the T-1 jack

Successfully looping your T-1 jack or the end of the cabling that leads to your CSU indicates that you have good physical continuity on your circuit to that point. If you are troubleshooting an issue where your carrier can't reach your

CSU, you probably have a defective CSU. If you had quality issues and the loopback tests came back clean to the hard loop on the RJ-45 jack, you again know that any static or frame slips you are experiencing aren't being generated by the wiring. Your next phase of testing is asking your hardware vendor to validate or replace your CSU, because that is the last possible variable causing your issue.

If you can't loop the T-1 jack

You are facing an interesting situation; you can loop the NIU but you can't loop the T-1 jack. There are only two variables between these two pieces of hardware:

- The first variable is the cabling that connects them.
- The second variable is the T-1 jack on your NIU into which that cabling is connected.

Follow these steps:

1. **Find your NIU (remember, it might be in a locked phone room somewhere else in your building).**
2. **Place your male loopback plug in the jack where the cabling for your circuit begins.**

   Depending on the results of the test, you have two scenarios:

   If your carrier can see the loopback plug and the NIU is good, the cabling is defective between it and your T-1 jack. In this case, you need to have the person who pulled that cabling dispatched to repair it.

   If your carrier can't see that loopback plug, the jack you are plugged into is defective and the local carrier needs to replace the NIU.

Step 5: Looping the CFA point

The *CFA point* is the physical point where your long-distance carrier connects into the local carrier's network. At this point in troubleshooting, you can't easily segment the circuit, so you need to determine whether the issue is with your carrier or with your local loop provider.

Figure 13-6 shows how the CFA point acts as the *demarcation point* (or *demarc*) between your carrier and the local loop provider.

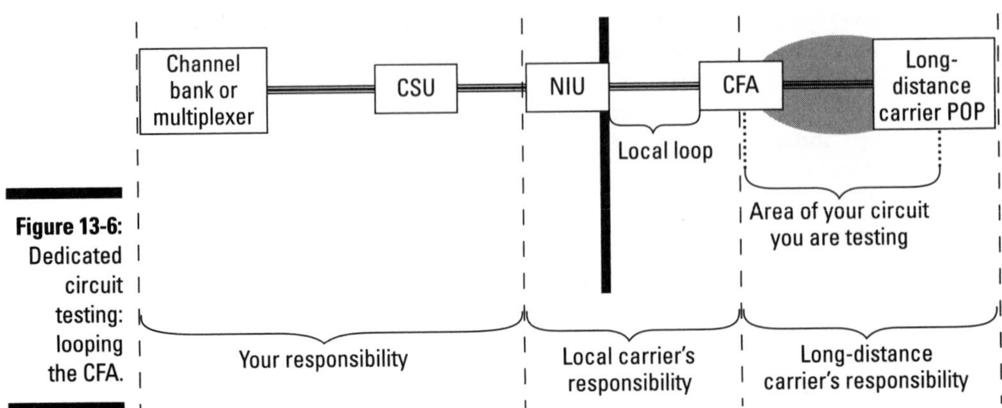

Figure 13-6: Dedicated circuit testing: looping the CFA.

If your carrier can loop its CFA point

If your carrier can loop the CFA point and run a clean test to it, but the carrier notes errors when running to the NIU, the problem is with the local carrier section of the circuit. If your long-distance carrier ordered the local loop, then one of its representatives has to open a trouble ticket with the local loop provider. (If you ordered the local loop, you have to open the trouble ticket on your own). This process can take four or more hours before the local loop provider responds to the problem and dispatches a technician. If the trouble ticket is opened late in the day, a technician might not be available until the following business day.

If you can't loop the CFA

If your carrier can't loop the CFA point, the issue lies within the carrier's section of the circuit. Because the problem is contained within one carrier, you could see a much faster resolution. Simply call the technician every hour or so to follow up until the situation is resolved.

Following a Dedicated Troubleshooting Shortcut

If your carrier asserts that a quality/continuity/protocol handshaking issue on your dedicated circuit is the result of problem within your hardware, there might be a very quick and easy way to confirm or deny that assertion. You can try this test only when you have two or more circuits with the carrier and only one of them is experiencing a problem. In this test, you unplug the cables that connect the CSUs to the NIUs and exchange them, as shown in Figure 13-7.

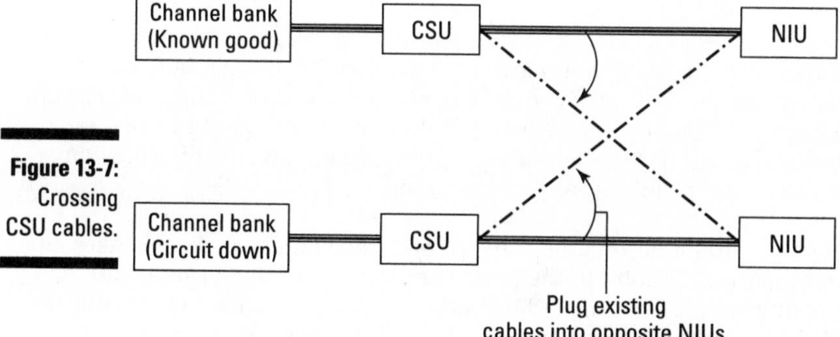

Figure 13-7: Crossing CSU cables.

After reconnecting the cables to match the diagram, reboot your hardware and assess the situation. If the problem ends up following the circuit from your carrier, and the multiplexer that was working fine is now failing or experiencing the issue, it's obvious that your carrier is the source of the problem. If the outage remains within the same multiplexer or CSU that was failing on the other circuit, you have a problem with your hardware.

This test is valid only when:

- Both circuits are configured exactly the same: Line coding/framing, outpulse signal/start
- Both circuits are on the same carrier (both the same long-distance carrier and the same local loop provider)
- Both multiplexers are the same make and model
- Both multiplexers are configured exactly the same (trunk groups, timer settings, and all the hardware specific settings)

If any of these conditions aren't met, the validity of this test disintegrates instantly. There are too many variations between carriers, configuration, and hardware manufacturers to say with conviction that if one set of your hardware works on a carrier, a different set should work as well.

Validating the Circuit You Are Testing

Incredible, but true! People have tested for hours on a circuit, with conflicting information coming from both the hardware vendor and the carrier, only to eventually find out each test performed was on a different circuit.

If your carrier sees the circuit up and you see it down, you may not be looking at the same circuit. If your carrier takes down the circuit, but you still see it running and passing calls, you definitely aren't looking at the same circuit. Having multiple carriers with several DS-1s per carrier might be great for your least-cost routing, but all those DS-1s can add a degree of confusion when you try to troubleshoot a single circuit.

Identifying a circuit is sometimes tough because you can make mistakes anywhere from the CFA point to the point the circuit enters your hardware. Confirm the circuit IDs assigned by both your long-distance carrier and the local loop provider. Then perform the following steps:

1. **Unplug your hardware from the T-1 jack or reboot your CSU.**

   Your carrier should see the circuit drop, bounce, or reset. If the carrier sees nothing happen on its end, the carrier needs to begin looking at the other circuits you have on its network. Perform the same test until everyone is sure you're dealing with the same circuit.

2. **Follow the cabling from your multiplexer to the T-1 jack and all the way to the NIU if you can.**

   If nothing else, the circuit ID of the span should be written on the T-1 jack in your phone room. Validate that the circuit ID matches what you have on file, and then proceed to the NIU for more investigation.

3. **Check the NIU to see whether you are connected to the wrong jack.**

   If the circuit you are working on was active and passing traffic at one time, this is probably not the issue, but if this circuit was recently installed, you may have discovered your problem.

4. **Check the circuit IDs listed on the NIU.**

   The NIU should list at least the local loop provider's circuit ID. These numbers should match the circuit ID listed on your T-1 jack in your phone room.

5. **Check the NIUs for changes when they are looped or sent into failure.**

   NIUs have lights on them that identify the disposition of the circuit they provide. A *loopback light* (LBK) is illuminated when your carrier is looping the NIU, and red lights are displayed when your carrier disables your circuit. You might need to check all of your NIUs before you find the one that reflects the work your carrier is doing on it, but this is the most foolproof method of identifying your circuit.

If your carrier performs all these tests and none of your circuits respond accordingly, the circuit you are working on either doesn't belong to that carrier, or is somehow absent from the carrier's visible inventory.

The Basics of Dedicated Toll-Free Troubleshooting

Before you begin troubleshooting your dedicated toll-free number, you must determine the extent of the problem. Opening a trouble ticket on your toll-free number will delay the resolution of your issue if the problem is actually with your entire circuit and also affects your outbound calling.

Just as with troubleshooting a switched circuit, it's always best to open a trouble ticket with your carrier for the outbound issue instead of taking a toll-free call that also has the problem and using that as the basis of your report. If your entire circuit is down, all of your toll-free numbers that terminate into that circuit will be affected. Check your circuit first, and only after you confirm that the problem doesn't affect your outbound calls should you open the ticket on your dedicated toll-free number.

The most common problems that plague dedicated toll-free numbers are routing issues and DNIS configuration issues. These are the only two variables that really separate dedicated toll-free numbers from dedicated outbound calls. They share the same dedicated circuit as your outbound calls and the same switched network as every other call you make.

The limited number of variables makes dedicated toll-free numbers relatively easy to troubleshoot. Quality issues that exist only on dedicated toll-free numbers are quite rare, because the dedicated circuit and routing they travel on are shared by outbound calls. The following sections only cover dedicated toll-free completion issues.

Step 1: Identifying a provisioning issue

Switched and dedicated toll-free numbers have inherent similarities. Either variety can die or lose its way during the process of being activated by your carrier. If a dedicated toll-free number is being migrated from another carrier, or is new and is slated to be activated, and the line suddenly fails, your first call is to the person at your carrier who handles your orders. You need to speak to this person first to determine whether the toll-free number has finished the provisioning process and whether the carrier sees it as active.

There are quite a few situations that can prevent a toll-free number from being activated. Until the number shows completed in your carrier's system, the customer service and troubleshooting people can't help you. Jump to Chapter 9 to find out how to resolve activation issues and for more information about the pitfalls and timelines for resolving provisioning issues.

Dedicated toll-free numbers must also be programmed in your PBX or phone system, especially if you use DNIS to route your inbound calls. If you made any upgrades to your phone system, the programming might have been lost. Check your phone system and confirm that all the routing is intact there before opening a ticket with your carrier. You might be able to find and resolve the problem yourself.

When you establish that the problem might be a provisioning issue, you need to contact the provisioner. The provisioner can confirm whether the toll-free number is still in the provisioning process. If the problem turns out to be a provisioning issue, you will have to work through the issue with the provisioner's assistance.

If you can't get ahold of the provisioner, do leave that person a voicemail message that explains the situation. Make sure you remember to leave your contact information. Then open a trouble ticket with the carrier's customer care department. You should have your provisioner check into the toll-free number when he or she picks up the voicemail, because you might save a lot of time. After you contact the provisioner, head to "Step 2: Redialing your dedicated toll-free number."

If your toll-free number had already been activated and was passing traffic prior to the problem cropping up, you can proceed directly to Step 2, as well.

Step 2: Redialing your dedicated toll-free number

All toll-free troubleshooting beings with trying to dial the number again. There is always the propensity to dial 800 rather than 888 or to transpose some digits. If your call completes during the second try, ring your number up again a few more times to make sure everything is fine before you continue with your day. If your call fails, write down the phone number you dialed from, the time you placed the call, and the call treatment you received. This information becomes your call example to share with your carrier when you open your trouble ticket. Next, move on to Step 3.

You can't dial a toll-free number from a dedicated circuit. This means that you can't dial out from your dedicated circuit to your toll-free number. When you are making test calls to a dedicated toll-free number, use a regular switched phone line (a nondedicated line) or a cellphone.

Step 3: Validating your dedicated RespOrg

There are two key elements involved in routing calls to a toll-free number: the RespOrg and the carrier receiving the traffic. Unless you have direct access to the national SMS database (which I cover in detail in Chapter 5), you can't confirm any of this information by yourself. Your carrier has access to this information. You must find out who controls the RespOrg status of your dedicated toll-free number.

If your toll-free number has not finished the provisioning process, you will be kicked back to the person who enters orders. Luckily, if you already left a message for your provisioner before you called (see "Step 1: Identifying a provisioning issue," earlier in this chapter) the carrier is already working on your issue by the time you call to follow up.

If your provisioner has already confirmed that the order is complete, don't let the customer service person off the phone! Request a conference call with the customer service agent and your provisioner so that the issue can be resolved immediately. It can take hours to get the service side and the provisioning side of a company to view the status of an order from matching perspectives. If your provisioner is unavailable and the problem is an emergency, ask the customer service rep to conference in the manager of provisioning. Don't be afraid to escalate the issue. As long as your issue is an emergency, nobody should fault you for pushing up the chain of command.

If your number was working fine yesterday, it might have been migrated or NASCed away. In either case, the RespOrg of your number may be some other carrier with which you don't have service. If the RespOrg of your number belongs to your carrier, proceed to "Step 4: Validating the DNIS configuration." If the customer service rep at your carrier doesn't recognize the company in RespOrg control of your toll-free number, you need to begin steps to reclaim your number. Your customer service representative should be able to tell you the following information about the company that has your toll-free number:

- The company's RespOrg ID code
- The company's contact phone number for reporting trouble

This is all you need to begin your work. Your carrier can't or won't do much in this instance without your direction. You need to call the other carrier and require it to identify why and how it received the RespOrg for your toll-free number. When you have that information, you can either leverage your position against the carrier if it NASCed or migrated the number in error, or plead for compassion if your carrier somehow released the number.

If your carrier released the number to the other carrier directly, or released it into the pool of nationally available toll-free numbers, you are between a rock and a hard place. Legally, the number belongs to the new carrier. The source of your problem is with your old carrier, which released the number. Taking legal action against your old carrier might make you feel good, but it does nothing to get your toll-free number back. You'll have better luck pleading your case to the new carrier. Let the new carrier know about all the wrong-number calls it is going to receive for the next six months because your customers will continue to call. And don't be too proud to beg.

Step 4: Validating the DNIS configuration

Most dedicated toll-free numbers are ordered with specific DNIS digits that are sent during the call setup portion of the call. The stream of information in which the DNIS is sent also includes the *caller ID* (what your carrier refers to as *ANI delivery*) and possibly a two-digit code that identifies the type of phone that initiated the toll-free call (such as a pay phone or prison phone), called *ANI Infodigits.* If you have questions on what DNIS is, browse Chapter 5. If you want to know more about how DNIS is ordered or provided, stroll to Chapter 9.

Provide the exact DNIS setup to your customer service rep for validation. If your toll-free number is 1-800-555-1234 and you ordered ANI delivery with a four-digit DNIS that matches the last four digits of the phone number, tell the agent that 800-555-1234 should have ANI delivery provided, and that the DNIS is 1234. The agent should be able to tell you whether ANI delivery is being sent, as well as the specific digits of the DNIS.

Dealing with DNIS digits that don't match your toll-free number

If your toll-free number was working fine yesterday, and suddenly the DNIS configuration has changed, your carrier should be able to fix it quickly. The carrier might have upgraded a switch and some of the programming might have been lost. Depending on how your carrier is set up, the customer service rep might be able to change the DNIS back and resolve your issue in moments. Otherwise, the rep may have to open a trouble ticket (curses!) and a technician will resolve the issue in due time. If the rep has to open a trouble ticket, take down the ticket number and follow up every hour or two until it's resolved.

Dealing with the situation if the DNIS digits do match

If the DNIS digits are correct, you need to press forward with a trouble ticket. Write down that trouble ticket number. Now you will have to wait for a technician to call you back before you can begin head-to-head testing in "Step 5: Head-to-head dedicated toll-free testing."

Using the old stare-and-compare method

If you have several dedicated toll-free numbers that are working fine and one that is failing, inform your carrier of this information and give the representative one of the working toll-free numbers to use as a template. If any change was made in the setup of your other toll-free number, it will be immediately obvious after the carrier compares the routing and configuration of the two numbers. Any variation in the setup of the two numbers is most likely the source of the completion problem.

This comparison works even better if you have another toll-free number that shares the same DNIS digits. If you have three toll-free numbers with the last four digits of 9876, all sending DNIS 9876, yet one of them is failing, a quick comparison with one of the functional toll-free numbers should quickly identify the problem.

Step 5: Head-to-head dedicated toll-free testing

Head-to-head testing on dedicated toll-free numbers is less painful than intrusive testing on a dedicated circuit. The process doesn't take down your dedicated circuit or affect your other service in any way. Your carrier can set up a test toll-free number in order to view and manipulate the DNIS stream, as well as to identify the responses from your hardware. After you get to the point of head-to-head testing on a dedicated toll-free number, your issue is as good as fixed.

Having your hardware vendor on-site or available during the head-to-head testing with your carrier can expedite the resolution. Your carrier can view and manipulate the call it's sending you, having someone with more experience around to identify what you are seeing can be infinitely useful.

Handling Dedicated Toll-Free Quality Issues

Static, echo, and low volume are examples of quality issues that sometimes occur on dedicated toll-free numbers; these issues aren't related to the dedicated circuit. Try to duplicate any of these issues on outbound dedicated calls to the same phone numbers that are trying to call your toll-free number.

I recommend trying the calls over a switched phone line to make the test even more effective and to help focus your carrier's repair efforts.

Generally, you can duplicate these quality issues and divert attention from both your circuit and your toll-free numbers, but sometimes this isn't an option. Say a carrier routes a dedicated toll-free call through one section of its network to arrive at the dedicated circuit without ever touching another end of the network. Conversely, a return call from the destination phone number to the originating dedicated toll-free phone number might take a totally different route back. For example, if you're calling from Charlotte, North Carolina, to a dedicated toll-free number in San Francisco, California, your call may be sent through the northern section of your carrier's network, running through Chicago, Illinois, and Seattle, Washington, before hitting the dedicated circuit in San Francisco.

This is in contrast with the route assigned through the southern portion of the network. A call from a switched phone line at your office in San Francisco back to Charlotte would run through Phoenix, Arizona; Dallas, Texas; and Miami, Florida, before hitting the network in Charlotte. Because these two calls don't share any similar routes, you can't replicate a problem on your toll-free number with an outbound call. In these rare instances, you need to push forward with resolution on your toll-free issues just as if they were switched quality issues. That means logging in all the affected and unaffected call examples you can lay your hands on, and relaying the information to your carrier. When the carrier has all the information, you simply have to wait and let the carrier do the job. If you become too frustrated, my only word of advice is to escalate the issue.

Part V

What's Hot (Or Just Geeky) in the Telecom World

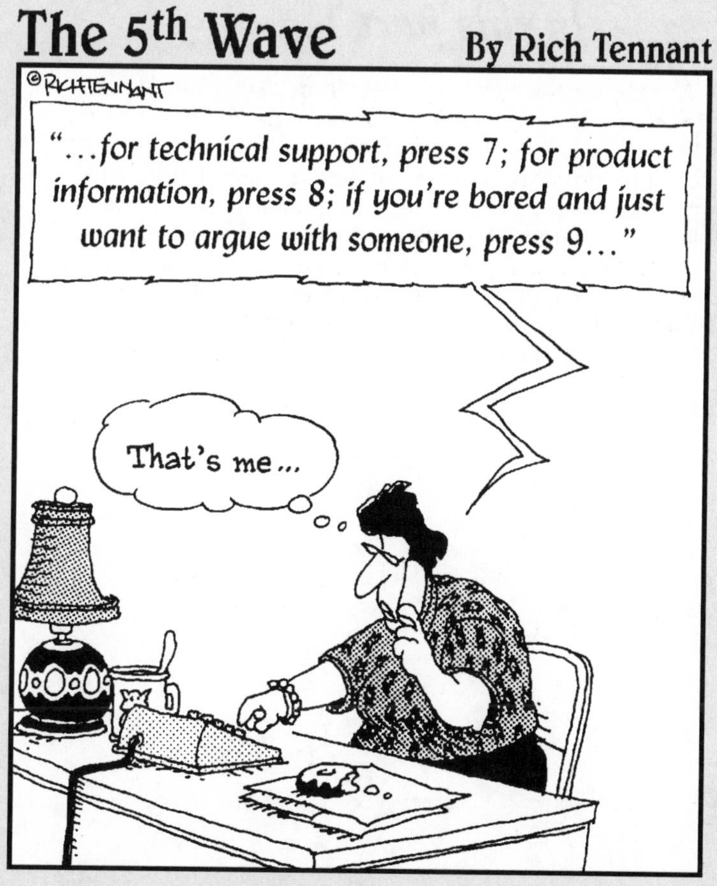

The 5th Wave
By Rich Tennant
©RICHTENNANT
"...for technical support, press 7; for product information, press 8; if you're bored and just want to argue with someone, press 9..."
That's me...

In this part . . .

Chapter 14 gives you a view from 20,000 feet of the most popular ways to transfer data and helps you find the best fit for your data needs. Although the chapter couldn't possibly break down data transfer methods to the molecular level (that's a whole *other* book), it does identify the items you will need to have if you want to use the technology, and also gives you a sense of the types of charges you can expect to pay. Chapter 15 introduces you to the makeup, features, and benefits of VoIP (Voice over IP), the hottest technology in telecom today.

Chapter 14

Transferring Data, Not Just Voice Content

In This Chapter

- Transferring data with modems
- Understanding the benefits and drawbacks of private lines
- Pricing a private line
- Using frame relay
- Understanding how a typical PVC is partitioned
- Using MPLS and its advantages over frame relay

Relaying information is the basis of all business. You have to manage your inventory coming and going, your cash flow, and all the fine details to prevent vital information from falling through the cracks. Fifty years ago, most information was passed along over a phone conversation or sent as a handwritten ledger through the U.S. postal service. At that time, high-speed *data transfer* would have meant sending mimeographed (or carbon-paper) copies of spreadsheets across town with your accountant.

Every company has data it needs to send within its company. The Internet offers more options for data transfer and greater access to information for customers and vendors. You can now make data downloadable in a PDF document, create interactive customer Web sites, and transfer large files quickly by using FTP sites. Of course, whenever data is being transferred, security is a huge priority and should always be factored into your data decisions.

This chapter covers the main forms of data transfer you will encounter, from modems to private lines, frame relay to MPLS. As you go through the options, you discover how each type of data transfer has its niche.

Understanding Your Data Transfer Requirements

You must ask yourself three important questions when you look at your data requirements are:

- **How much data do you need to send and how quickly do you need it?** Your answer to this question will frame the parameters of the next two questions. Say, for example, you have an inventory system containing 10 million items that are housed in 3 different warehouses and are sold from 500 retail stores across the globe. You might need an accurate listing of the entire inventory available to every one of your retail stores, in real time. On the other hand, you might only send one large accounting file to your main office once a week. The amount of data you need to send shapes your choices for how you will send it.
- **How many locations are involved in your data transfer?** As the number of sites you need to share data with increases, so do your options. You can easily and cost effectively send data between two locations, but the methods you might use in such a scenario become prohibitive if you have more than three locations. The number of locations involved, in addition to the quantity of data you need to send, determines your exact data requirements.

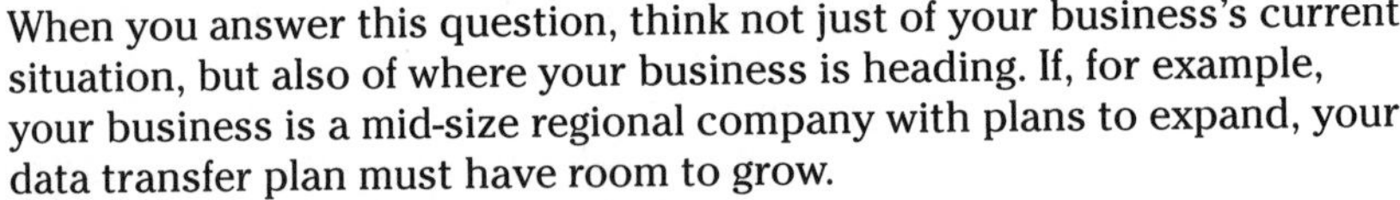

When you answer this question, think not just of your business's current situation, but also of where your business is heading. If, for example, your business is a mid-size regional company with plans to expand, your data transfer plan must have room to grow.

- **How much are you willing to pay to make this happen?** You can pay as much or as little as you want to transfer your data. Overbudgeting your data transfer network might make everything move very quickly, but it can also be very expensive. Telecom salespeople who aren't familiar with data applications often quote a wide range of capacity options, hoping to offer one that fits your budget. The problem with this approach is that you can easily end up buying too little bandwidth, simply based on the price. The more money you pay, the more bandwidth you should get. But, do you need all that bandwidth?

Transmitting Data the Old-Fashioned Way

Most data is transmitted by simply attaching a document to an e-mail message and clicking the Send button. About 20 years ago, the same transfer might have been made using a direct connection into the office server and sent via modem. Modems do transmit and at very slow speeds when compared to other forms of data transfer, but still exist today and are best used in two niche markets:

- **Administration applications:** Phone systems and local area networks aren't infallible. If your LAN is failing, you might not be able to access your data. However, a technician can still dial into the router with a modem and troubleshoot the system. The same scenario goes for dialing into a PBX or channel bank for troubleshooting. As long as you are dialing from your computer directly into your server or PBX, you have a secure connection for as long as you need it.
- **Older systems that require modem transfers:** Not everyone has jumped on the DSL, or cable modem, or T-1 bandwagon for their Internet connection, and some large software systems were designed to use modems to transfer data (such as many HR programs). In spite of the speed at which technology advances, many systems and people have not upgraded to faster connections.

Aside from these two niche applications, there is little reason to use dialup modems to transfer your data unless you like waiting.

Processing Constant Transmissions between Locations

Setting up a *private line,* sometimes called a *leased line*, means actually renting the copper wire that connects two offices. If you use a long-distance modem connection for eight hours a day, five days a week, in order to access some information, it might be more cost-effective to simply rent the wire in between the offices instead of paying the per-minute charge for modem transfers.

Businesses haven't been dialing up with modems to transfer large amounts of data for a while, so the primary use of leased lines today is to act as an extension of your local area network (LAN). You can order a private line as large as you want, from a T-1 line capable of passing 1.54 Mbps (megabits per second); you can also order a DS-3 line, which can pass 45 Mbps, or an OC-3 line, which passes 155 Mbps. When you have the private line built, it is little more than a large empty pipe. This fact gives you a couple of areas of flexibility:

- **You can transmit any kind of data you want.** A private line gives you a link between two locations that can be used for anything from voice phone calls to videoconferencing and data transfers.
- **You can transmit data in whatever method that works for your network.** You can transmit voice calls via normal TDM (time division multiplexing) technology, or maybe you want to use Voice over IP (VoIP) technology.

When you order your private line, your carrier will identify each end of the circuit as either the *A location* or the *Z location*.

Private line pricing

Private line pricing increases as the distance between the two offices increases. The size of the private line you request also determines pricing. The price you receive for a private line includes the cost for local loops from both ends to the nearest Point of Presence (POP) for your long distance carrier and the long-haul mileage between the POPs.

The long-haul cost is generally based on a specific charge per mile per DS0. If you have a 100-mile long haul for a T-1 circuit where the per-mile charge was \$0.20, you would expect this section of your circuit to cost \$480.00 (0.20 x 24 channels in a T-1 x 100 miles). If the circuit was a DS-3 rather than a T-1 the long-haul portion alone would be \$13,440.00.

Figure 14-1 shows how a private line is built with two local loops that span from the long-distance POP to the end locations, with the long-haul mileage in the center. Because private line quotes only list two items, the installation fee and the monthly charge, you won't know how these totals are derived. If the quote is excessively high, ask for quotes from other carriers. The local loop fees vary from one carrier to another depending upon where their POPs are located in relation to your office. The difference in this one element alone could save you hundreds of dollars a month.

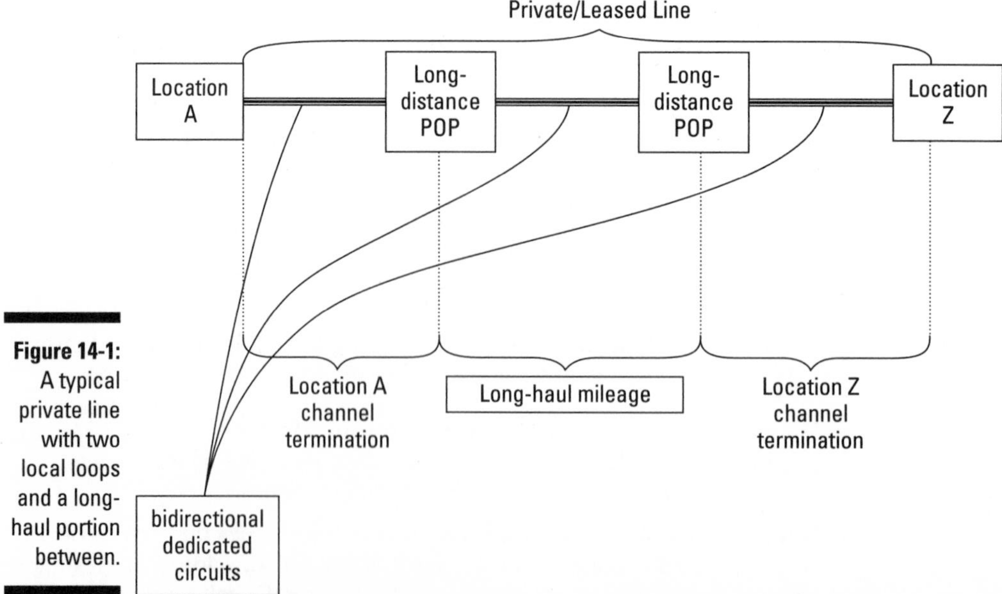

Figure 14-1: A typical private line with two local loops and a long-haul portion between.

Private lines are a low-profit item for most carriers, so they often encourage you to sign a longer contract by reducing the cost as the terms increase. The difference between a 12- and a 36-month contract might be substantial, but remember that you are on the hook for 2 more years than you might want to be. The one guarantee I give you about telecom is that it's always changing. Any contract longer than 12 months might prevent you from upgrading to a newer, better, less expensive alternative.

Drawbacks of private lines

The downside of private lines is their cost, and the fact that each private line can connect only two sites. The farther a private line stretches, the more it costs per month. A T-1 private line from Los Angeles to New York may cost $6,000 per month, and if you don't use it very often, it may not be worth the monthly payment. Nowadays, businesses can send data in e-mails or across the Internet. If you deem the level of security on the Internet acceptable for your purposes, private lines might not be such a great deal.

The other problem with private lines occurs if you want to connect more than two locations. You *can* connect more than two locations with private lines, but this setup is anything but cost-effective. Figure 14-2 shows what it would take to tie together offices in Seattle, Washington; San Diego, California; Miami, Florida; and Boston, Massachusetts. Each line that spans from one POP to another is a private line, with a total of six lines required to provide complete redundancy. This type of setup is geared for security and invariably builds a network with too much capacity, and a monthly cost that could easily top $50,000. Yikes!

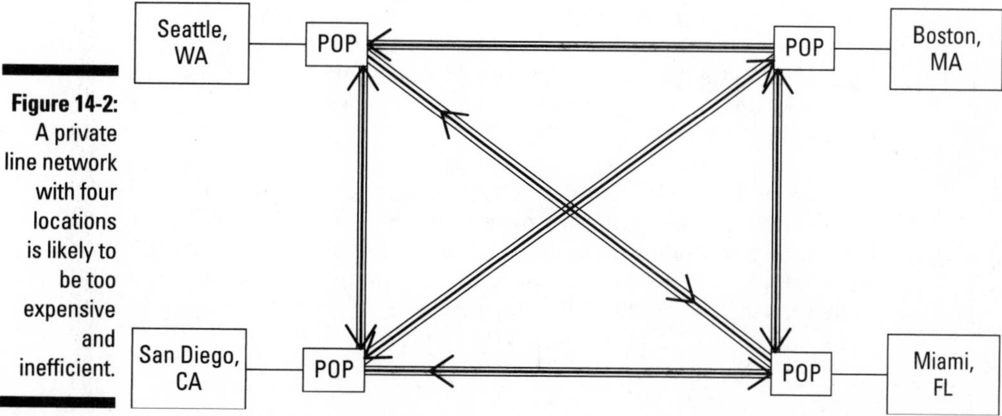

Figure 14-2: A private line network with four locations is likely to be too expensive and inefficient.

Understanding a Frame Relay Network

Frame relay is the next generation of data transfer solutions, and it was created to replace private line networks that used to crisscross the nation, costing businesses too much money. Sometimes private lines are installed for redundancy, but they are rarely used to their full potential because they exist only as a safety net if another connection fails.

Frame relay takes information from the office, stuffs it in a nice little package, places a destination and return address on it, and sends it on its way along with all the other packages the carrier is processing. Figure 14-3 shows a typical, fully redundant setup for a frame relay network with four *nodes* (individual locations on the frame relay network).

Frame relay is a logical transition if you currently connect multiple offices with private lines. It is also an option if you're looking for a more secure way to transmit data to many offices, because frame relay is more secure than the public Internet.

Drawbacks of frame relay networks

One of the drawbacks of frame relay is that you must order dedicated circuits to connect each of your offices to your carrier's frame relay network. These dedicated circuits are shown as the triple lines in Figure 14-3, connecting each location to a POP; they typically cost between $200 and $800 each for a T-1 connection.

Frame relay technology was a great step forward when compared to private lines. However, the cost of PVCs (permanent virtual circuits), and the limited number of PVCs available, are issues that you can't ignore. For more information about PVCs (which are represented in Figure 14-3 with dashed lines), check out the following section, "Frame relay element #3: The permanent virtual circuit, or PVC."

In addition to the PVC problem, frame relay networks are unable to route calls around problem areas or prioritize them. These issues in particular have spawned a new technology called *Multi-Protocol Label Switching*, or *MPLS.* MPLS was designed to overcome the limitations of frame relay. For more information about MPLS, see the section, later in this chapter, called "New and Improved Transmission for Multiple Locations."

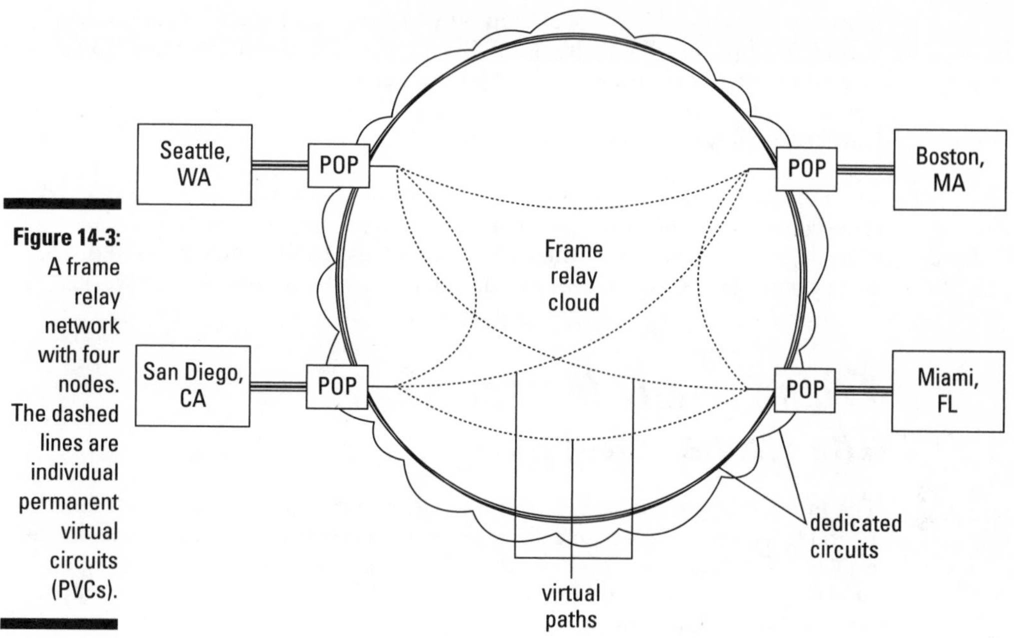

Figure 14-3: A frame relay network with four nodes. The dashed lines are individual permanent virtual circuits (PVCs).

The *frame relay cloud* in the center of Figure 14-3 makes setting up data transfer between four distant locations more cost-effective than the same setup with private lines. As you see, all the lines within the cloud are dashed, indicating that the path your data travels isn't a dedicated span that only you alone can use, but is instead a virtual path between two points that many customers share.

Frame relay has three main pieces that allow it to work, and each one is generally listed as a separate line item when you receive a quote for the service from your carrier. These three charges apply to every site connected to the frame relay network, and if any are missing when you receive your quote, you should confirm that the locations are available through your carrier.

You need to have some hardware if you're going to have a frame relay network. The following sections list all the elements you need to order. The elements are available in a variety of sizes to fit your exact requirements.

Frame relay element #1: The local loop

Just like the long-distance circuit needed for voice calls, you have to order a *local loop* to connect the router at your office to your carrier's frame relay network. You can order a local loop only in standard dedicated loop increments: 64 kbps, full T-1, or full DS-3. Even if you only need 256 or 512 kbps, you still have to order a full T-1 for your local loop. Your carrier might tell you that it can provide a *fractional T-1* (a portion of a full T-1) local loop, but it will still charge you for the full bandwidth.

The cost of the frame relay local loop might be more than a local loop used for a voice circuit, because your carrier has to receive the circuit in at a certain place in the network where the carrier has access to the frame relay network. Not every voice POP your carrier has can access the frame relay network.

Frame relay element #2: The frame relay port

The *frame relay port* is where your local loop terminates at the entrance portal to your carrier's frame relay network. The port can vary in size, based on 64 kbps increments (just like the DS-0 channels on a T-1), depending on how much bandwidth you need. This means that you can order a 256 kbps port, but not a 220 kbps port.

Frame relay element #3: The permanent virtual circuit, or PVC

The *permanent virtual circuit,* or *PVC,* is the dashed line within the frame relay cloud that allows your data to travel from one frame relay node to another. Your data doesn't have exclusive rights to the path it takes between the nodes as it does in a private line. Your data is instead packaged into a frame, numbered sequentially based on how it was made, and labeled with a destination address in the header before it's sent off into the frame cloud. When it's in the frame relay network, your data is routed along to the required destination at the other end of your PVC.

You can purchase the PVC, just like the port and the local loop, in varying sizes, depending on your requirements. Just like the frame relay port, you can request the PVC in any standard telecom size in increments of 64 kbps (sometimes even in 56 kbps increments).

PVCs are identified within the frame relay network by *data-link connection identifiers (DLCIs),* which the frame relay provider assigns to each end of the PVC. These DLCI numbers exist in a limited range from roughly 16 to 991 (the available range of numbers has changed over time) and are only recognized within the frame relay network. This limitation was overcome by an enhancement including *Local Management Interface (LMI)* functionality, thus allowing additional nodes to be attributed to a single DLCI known within the frame relay network.

When you order a PVC, you are rarely given all the bandwidth you ordered. When you order a 256 kbps PVC, for example, you might be guaranteed only a portion of that bandwidth. Figure 14-4 gives you a good idea of how a PVC is partitioned:

512 kbps PVC
The final 64kbps is listed as your Excess Burst Size, or B_e.
An additional 192kbps is listed as your Committed Burst Rate, or B_c.
256kbps is listed as your Committed Information Rate, or CIR.

Figure 14-4: How a typical permanent virtual circuit (PVC) is partitioned.

Understanding your virtual circuit's bandwidth needs

Your frame relay network's permanent virtual circuit (PVC) is actually composed of several layers of bandwidth that you don't have complete and constant access to. The percentages of your total PVC that make up each layer are sometimes set by your carrier, or you might be able to adjust the percentages by paying more money for larger sections of guaranteed bandwidth. The following sections identify your options.

Committed information rate: CIR

The *committed information rate*, or *CIR*, is the most important part of your PVC. This is the amount of bandwidth you have guaranteed access to at all times.

CIR sizes vary by carrier. Every carrier has a different policy on how it assigns the CIR for its PVCs. Some carriers give you a CIR that is equal to the size of the PVC you have ordered, and some guarantee you only a quarter of the PVC bandwidth for your CIR. You might be able to pay more for a higher level of CIR, so check with your carrier if necessary. As a standard rule, request at least 50 percent of the total bandwidth of your PVC to be designated as

CIR; you don't want to lose data because you are using too much of your bursting bandwidth.

Committed burst rate: B_c

The *committed burst rate* is the additional bandwidth your carrier guarantees you can burst to for short duration use. The carrier generally doesn't tell you how long you can burst and use this bandwidth; it only tells you that it's for bursting. This bandwidth is used during your peak times when you are sending or receiving large quantities of data, but isn't the average bandwidth used during normal business hours. This extra burst of bandwidth is for instances such as sending your weekly database file to the accounting office. It isn't the typical transmission level you work at on a day-to-day basis, but once a week you may need the extra room.

Excess Burst Size: B_e

The *excess burst size* is the additional bandwidth that you can use, as long as nobody else is using it. Imagine that two PVCs built next to each other are five lanes of a highway, with two lanes devoted to traffic in each direction and the shared center lane to be used for passing. The center lane is your B_e bandwidth.

B_e is room that you can use, as long as nobody (that is, no data) is coming in the opposite direction. The router sending packets into the PVC from your side can monitor traffic it is sending, but can't anticipate whether another router in the frame relay network is sending any large files back over the PVC. If you have two packets of data come into contact in the B_e bandwidth section of the PVC, both frames of data are blown away and sent into the bit bucket. Luckily, the data is automatically sent again, so you don't really lose that much.

New and Improved Transmission for Multiple Locations

Multi-Protocol Label Switching, or *MPLS,* is a technology designed to overcome the limitations of frame relay technology. To fully understand MPLS, you must understand a tiny bit about networking standards. The seven layer *Open System Interconnection Reference Model (OSI model)* is the networking standard that network engineers conform to when building networks. This model creates a unique division of labor both within and between each of the seven layers. Our discussion of MPLS only concerns the first three layers of the OSI model, sometimes called the *lower layers.* The *upper layers* of the OSI model aren't covered in this book. If you're interested in knowing how networks work, check out *Networking For Dummies,* 7th Edition, by Doug Lowe (Wiley).

The seven layers of the OSI model are numbered from bottom to top, so the lower layers are labeled like this:

- **Layer 1: Physical Layer:** This is the physical medium used to transmit data across the network. Literally, it's the cable through which data moves; it could be made of Cat 5 (Ethernet) cable, coaxial (coax) cable, twisted-pair shielded copper wire, or fiberoptic cable.
- **Layer 2: Data Link Layer:** This is the lowest layer at which software protocols work. The protocols that operate at this level include Ethernet, frame relay, and ATM (which is kind of like frame relay — on steroids). These protocols have relatively simple live data transfer schemes. Data packets are sent through the network on predetermined routes from one end of the network to the other. Because these paths are predetermined, the hardware doesn't have to decide the best path; data that is sent to a specific DLCI on a frame connection can only be sent to the part of the PVC to which the DLCI is assigned.
- **Layer 3: Network Layer:** The hardware at this level is more intelligent and is where MPLS shines. MPLS enables carriers to create *virtual private networks* (VPNs) similar to the frame relay networks of old. The increased sophistication of this technology allows it much greater functionality, and gives the network more resilience as it interacts at the Layer 2 and Layer 3 level for specific tasks.

Understanding why smarter is better

MPLS enables all the benefits of Layer 3 routing with the speed and security of Layer 2 switching. The flexibility of MPLS allows it to interact with and easily route data processed via Layer 2 and Layer 3 protocols.

The main benefit of having the Layer 2 protocols available with MPLS is that you can keep your existing equipment, which uses old protocols and transports data between sites with some of the more advanced routing capability available only via Internet Protocol. An MPLS VPN allows you to connect your company's old frame relay switches together or use Ethernet to tie your new corporate e-mail server to T-1s at your remote office locations. The packets in your MPLS VPN can also be prioritized so that your VoIP packets are routed before the accounting file being sent to your boss. This capability to avoid congestion and prioritize the different forms of data you are transmitting, while still maintaining a high degree of security, makes MPLS a very attractive data transfer method.

Figure 14-5 shows a standard setup of an MPLS network. Although MPLS VPNs are different than basic IP, ATM, and frame relay networks, as a customer, you don't need to deal with those issues. Your IT department can continue to refer to the different sites on your network via IP addresses if your organization had a standard IP network before. MPLS gives you the benefit of IP with the comfort of the network your IT department is used to. The interesting aspect to note is that there are no preset routes between any of the endpoints. Because MPLS functions at a network level, the routers within your carrier's network can receive data packets and direct those packets along dynamic paths to their final destinations.

The components that make up an MPLS network are similar to the pieces that make up a frame relay network.

Label Edge Router: LER

The *Label Edge Router,* or LER, is an important piece of hardware generally provided by your carrier, and is necessary to establish an MPLS network. The LER typically sits on the edge of your carrier's network. Your router is connected to the local loop that connects to the LER at your carrier. The main job of the LER is to assign a label to every packet of data sent through the MPLS network. This label determines the packet's routing process and priority.

The following sections explain the materials you need to have in place to get your MPLS configuration off the ground.

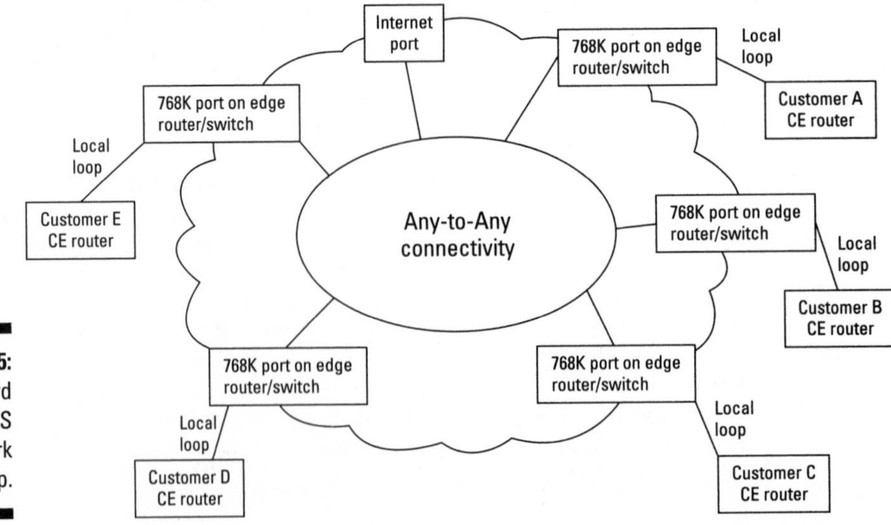

Figure 14-5: A standard MPLS network setup.

The local loop

Just like any dedicated service, you need to rent the copper wires or fiberoptic cable (depending on the size of circuit you require) from your office to the nearest MPLS entry point for your carrier. The local loop's cost is similar to that of a voice circuit of the same bandwidth, with the price increasing as the distance grows.

The MPLS port

Your carrier's Label Edge Router (LER) sits on the other side of your local loop to receive your data. As the carrier's LER receives each packet, it reads its label and routes it accordingly. You have to pay for a port into the carrier's LER based on your required bandwidth.

Bandwidth

The last item you have to pay your MPLS carrier for is the bandwidth required to send data. Carriers charge for bandwidth in a variety of ways, so speak to your carrier to decide what is the best billing choice for your specific application.

Researching MPLS on the Web

If you are interested in finding out more about MPLS before you call your network technician to talk over the project, you can tap the following Web resources for general information:

`www.juniper.net/solutions/vpn`

`www.globalcrossing.com/xml/services/serv_data_ipvpn_over.xml`

`www.cse.iitk.ac.in/users/braman/courses/cs625-fall2003/lec-notes/lec-notes32-4.html`

For Cisco troubleshooting, see:

`www.cisco.com/en/US/tech/tk436/tk428/technologies_tech_note09186a0080094b4e.shtml`

Chapter 15

Riding the Internet Wave: VoIP

In this chapter

- Understanding what VoIP does
- Getting a handle on IP protocols and codecs
- Setting up your business with VoIP
- Getting the lowdown on VoIP hardware
- Testing Interoperability
- Transferring local numbers

Voice over Internet Protocol, more commonly referred to as *VoIP,* is the hottest buzzword in telecom today. The downside of being on the cutting edge, however, is that VoIP is still finding its place in the world. VoIP technology has to be built into the infrastructure, and that takes money and long-term planning. In many ways, the world just isn't ready for VoIP, because it creates new problems even though it successfully solves older ones.

This chapter covers the general environment of VoIP, its structures and requirements. Figuring out how to program your Cisco router to convert VoIP to Feature Group D protocol is not covered (so it's cool if you don't know what I just said right there), but you do discover VoIP's greatest strength — its flexibility. The possible applications of VoIP are rapidly growing. The technology is used with everything from interactive Web sites to *find-me-follow-me* phone services that allow one call to attempt to reach you at several phone numbers before it is finally sent to voicemail. The ability to reinvite calls to new destinations: your cellphone, your office number, your home phone number, is one of the great benefits of VoIP, along with the limited government regulation and taxing at this time.

If you want a more in-depth look at VoIP that's written for real people, check out *VoIP For Dummies* by Timothy V. Kelly (Wiley).

Understanding VoIP Basics

At its most basic level, VoIP is a technology that allows you to place a voice phone call over an Internet connection. It uses an Internet connection to pass voice data, using the *Internet Protocol* (IP) rather than using *POTS* (plain old telephone service).

VoIP's applications are endless. Remote employees can work just as if they were hooked into your telephone system in-house, and because you're using your Internet Service Provider (ISP) to make VoIP calls, you're not getting charged by your long-distance carrier. Even though high-end VoIP setups cost a bundle to install, the technology ultimately keeps corporate telephone costs low, as long as your network is configured just right.

VoIP technology was initially used by modern-day versions of ham radio operators and computer geeks, who wanted to make free phone calls with poor quality over dialup Internet connections (remember dialup?). The processes and sophistication of VoIP has grown exponentially over the years, to the point where most long-distance carriers now use VoIP technology to transfer calls within their own networks, and large businesses are adopting this newer, higher-quality VoIP for their day-to-day operations.

The VoIP protocol not only converts your words into digital code, but also puts that code into packets for transmission across an Internet connection (this process is referred to as *packetization).* Your call eventually connects to another VoIP phone, or a piece of hardware called a *gateway.* At that point, the call is translated from VoIP to TDM (time division multiplexing protocol), a standard non-VoIP protocol that you can read more about in Chapter 8. From there, it's sent on to the destination phone. Gateways bridge the VoIP and Non-VoIP worlds together by translating the protocols, allowing your call to reach your grandma's rotary-dial phone — which is definitely *not* using VoIP.

Figure 15-1 shows how a call originates from your phone as VoIP and is transmitted along your Internet connection until it is converted to TDM with a VoIP gateway. Your call emerges from the gateway into the Public Switched Telephone Network, or PSTN, and is routed to the phone you dialed. Read the figure from right to left, rather than left to right, to see how calls originating from non-VoIP phones are received with VoIP.

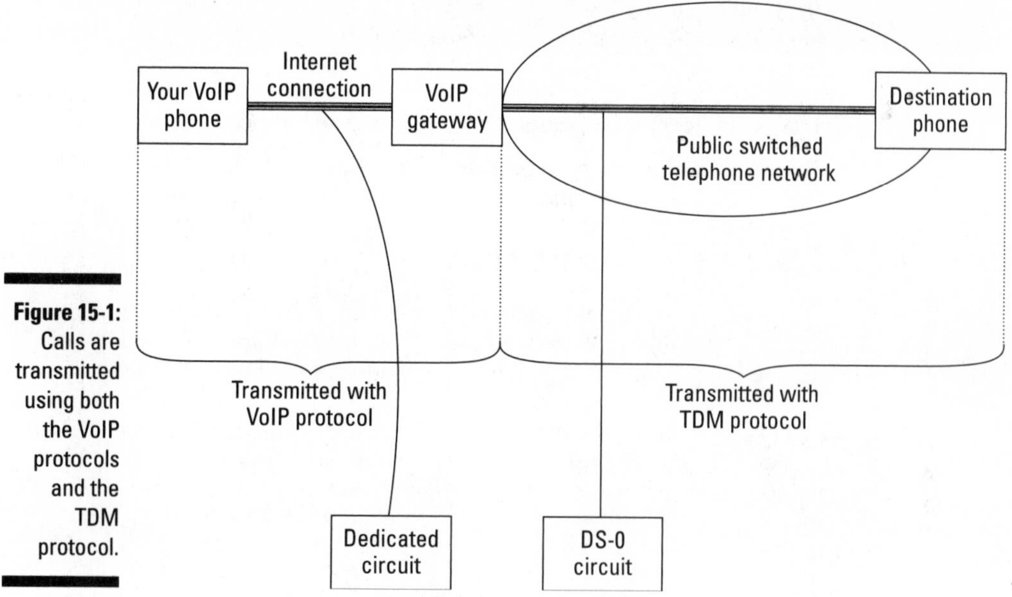

Figure 15-1: Calls are transmitted using both the VoIP protocols and the TDM protocol.

Understanding IP Protocols for VoIP

VoIP protocols come in several flavors, but two of them lead the pack in popularity. The *H.323* protocol was established by the International Telecommunications Union (ITU), and is geared for multimedia applications. The *Session Initiation Protocol* (more commonly known as *SIP*) was established by the Internet Engineering Task Force (IETF), and is quickly growing as the preferred protocol of choice with many carriers. Both of these protocols are used as a basis for transmitting VoIP calls and use the same coder-decoders, such as G.711 or G.729, to transmit and receive the calls.

Comparing H.323 and SIP

Despite the fact that its name isn't very user friendly, H.323 is the incumbent protocol, gaining initial market share over SIP. In fact, the name isn't the only part of the protocol that's hard to understand. Because H.323 is so complex, many companies have turned to SIP, which is an easier protocol to administer and manage. Looking at H.323 against SIP, we see that:

- **SIP is structurally faster than H.323.** It uses less time setting up calls, requiring only one invite message, versus the eight messages for H.323. Speed of call setup is an important factor for companies that use VoIP for telemarketing campaigns. If you place hundreds of thousands of calls a day, and every call means revenue, every second counts.
- **SIP is made of an easy-to-read text-based protocol.** H.323 is based on a binary software code that is difficult to read and comprehend. The text-based nature of SIP enables it to be easily dissected and analyzed, directly as it is transmitted from your server.

If you have a choice when choosing a VoIP setup, then I recommend using SIP over H.323. The industry is moving away from binary-based codes, so if you were to use H.323, you might need to upgrade sooner rather than later.

Ethereal is an extremely helpful tool for troubleshooting VoIP issues, allowing you to see every message sent through the hardware it is installed on. This information includes the IP addresses being used for transmission, as well as the call setup and teardown for easy troubleshooting. The software is available for free at `www.ethereal.com`.

Understanding the structure of a VoIP call

Every VoIP call consists of four completely separate data streams, each governed by the parent protocol, which is generally either H.323 or SIP (see "Comparing H.323 and SIP"). You have four data streams, established in pairs, with one pair at each end of the call.

One transmission set handles the setup, maintenance, and teardown (signaling) of the call, and the other set concerns itself with only the transport of voice data. Here are the details:

- **One pair of data streams handles signaling data to establish, maintain, and tear down the call.** H.245 is the most commonly used protocol for H.323 and *Session Description Protocol (SDP)* is the favorite with SIP. These protocols were designed to negotiate the signaling of a voice call over VoIP. There are differences in how the signaling is managed in SIP versus H.323, but by design, they are structured to keep the signaling removed from the data stream that holds the voice portion of the call.
- **The other set of data streams sends the voice data.** The Real-Time Transport Protocol (RTP) is the means by which your voice or video is sent from the originating phone or videoconferencing system to the system receiving the call. The signaling stream tells the VoIP equipment how to package the data and converts it on the far end of the call. RTP is the carrier pigeon that takes your message along the path determined by the signaling protocols.

Reading in the Ethereal world

Ethereal is a software package that enables your VoIP switch to capture packets, as long as the switch is using an OS *(operating system)* into which Ethereal can be installed (Ethereal runs on all popular platforms, including Windows, Unix, and Linux systems).

Ethereal only captures packets that flow through the computer on which it is installed, so if you have Ethereal loaded on a server behind your router, you will only be able to record packets passed between your router to your IP phones, and not the packets between your router and your carrier.

Figure 15-2 shows the four paths taken in a standard VoIP call. If any one of the connections doesn't terminate to the correct location, the call is impaired. Terminating the signaling stream to the wrong IP address on an unknown router prevents the call from being established, and terminating the RTP at the wrong location either prevents any sound from being transmitted, or results in one-way audio.

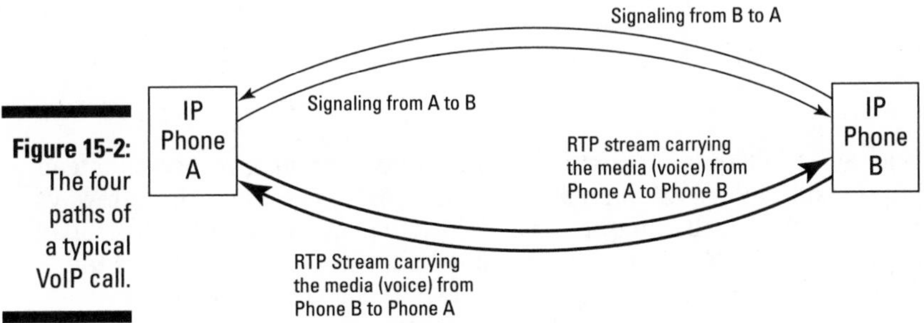

Figure 15-2: The four paths of a typical VoIP call.

Figure 15-3 demonstrates the VoIP's flexibility: When signaling and RTP streams can operate independently, you have a whole lot of options. In the figure, one phone (IP Phone A) is calling another phone (IP Phone B). The RTP stream is being reinvited to the media server that provides voicemail for IP Phone B.

The interesting thing about this call is that the media portion of the call (that is, the voice part) in the RTP stream is being sent to a media server, and the signaling protocol doesn't establish a connection between either of the phones and the media server.

You can configure VoIP protocols to try several different phones (a cellphone, another office phone, or a home phone). For example, say you decide that after five rings, if the call isn't answered, the call is sent to the next phone on the list. Such features are commonly referred to as *reinvite features*.

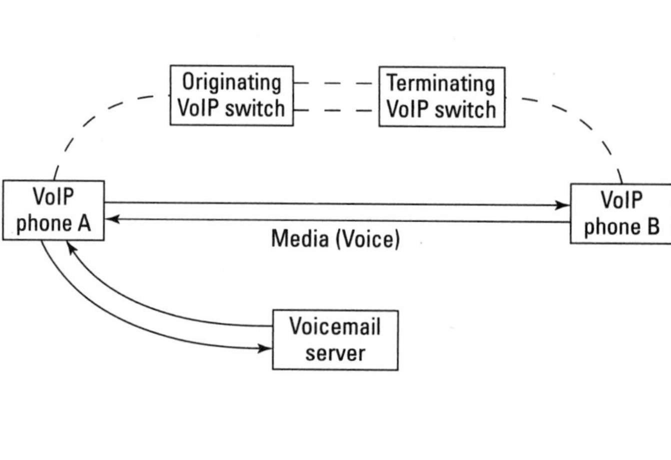

Figure 15-3: The voice portion of this call in the RTP stream is being sent to a media server. The signaling protocol never connects between either of the phones and the media server.

Data signals aren't sent in constant streams; the signaling protocols instead tend to transmit information on an *as-needed* basis. As long as the gateways at either end of the call function properly, there is no need for the overhead to baby-sit the call. After the call is established, the signaling handles the maintenance and housekeeping of the call while it is active, but doesn't transmit nearly as much information as is sent in the RTP stream.

Don't go rogue on me

If the SDP portion of the call is incorrect, someone may end up sending the RTP stream to an unknown IP address. A breakaway stream or *rogue RTP* is sent to some unlucky IP address in the world. The constant barrage of data may not be dangerous to the server receiving the rogue RTP, but it can be annoying. Performing interoperability tests with your carrier is crucial to working out configuration issues that would otherwise make your calls go to dead air.

Understanding your voice choices

Both H.323 and SIP protocols use the same equipment called *codecs* (an abbreviated name for *co*der-*dec*oder) to convert your voice into digital code so it can be transmitted. Codecs are built into Analog Telephone Adapters (ATAs), such as the ones used with Internet VoIP services like Vonage, and in VoIP gateways your standard phone service may connect to with your VoIP carrier.

Codecs use industry standard methods to convert voice signals into digital data. Converting your voice to digital code is a delicate process, with three primary concerns with data transmissions:

- **Compression ratio:** This ratio identifies how much a codec can reduce the volume of data sent to transmit the RTP stream of a VoIP call. The goal is reducing the amount of bandwidth you use to transmit your VoIP calls, enabling you to send more calls over the same bandwidth.
- **Call quality:** Calls can suffer from clipping, jitter, and latency to greater or lesser degrees, depending on the compression ratio and packetization process.
- **Packetization delay:** When a call is converted to or from VoIP, the process causes a delay. The length of this delay affects call quality.

As you can see, these three elements are inextricably related. All codecs used affect call quality and generate a *packetization delay.* As a result, when it comes to choosing a codec, your options are much limited based on your bandwidth and call quality requirements.

After comparing these elements, and looking at all the codecs, a few protocols stand out and are more commonly used than the other by carriers. I discuss them in the following sections.

Using uncompressed VoIP with G.711

G.711 is the garden variety, uncompressed protocol for SIP. If you want to send faxes or have poor line quality on another SIP protocol, G.711 is your codec of choice. This codec protocol gives you the highest quality and is most common in the industry.

This codec is the easiest to use, as the software is widely available on the Internet for free and all carriers support it.

G.711 comes in both a domestic U.S. version and a European version. The difference between the versions is minimal, but they're different enough to prevent them from working together efficiently. The version use in the U.S. and Japan is called G.711μ (referred to as *G.711 mu-law,* or sometimes called *u-law*) and the European version, which is called G.711a (often referred to as *G.711 a-law*).

The greatest drawback to this codec is the fact that, because the data isn't compressed, it uses up more bandwidth to transmit one voice call than you would use if you sent it out using the traditional Time Division Multiplexing (TDM) protocol. A non-VoIP call uses 64 kbps to transmit a single conversation, where the G.711 codec uses about 84 kbps. If you use this codec, your T-1 line, which in a non-VoIP scenario could handle 24 consecutive calls, is maxed out at about 18 consecutive calls. Is this the price you pay for quality? Maybe the benefits of VoIP outweigh the need for extra bandwidth, but you should definitely bear this issue in mind.

Compressed VoIP with G.729

VoIP with the G.729 is a 5:1 compression codec that allows VoIP to shine. It takes the 84 kbps required to handle a call with G.711 and reduces it down to only 20 kbps. In spite of the fact that the signaling portion of the call is not compressed, the voice portion is actually compressed to a ratio of 8:1. You can make almost 77 calls on the same T-1 of bandwidth where you were restricted to 18 with the uncompressed codec.

Faxes don't work when sent over the G.729. The codec was designed to compress the audio portion for voice calls by removing the inaudible tones. Unfortunately, inaudible tones are what make fax machines work. The tones are used to transmit data over the wires. If the inaudible tones aren't transmitted, faxes are doomed to failure on this codec. Don't fear; if you need to send compressed faxes, the next codec is for you.

Compressing your fax calls

The T.38 codec was built to resolve the faxing problem found on G.729. The T.38 *Fax over IP* (FoIP) protocol (oh, yes, there's a cute acronym for everything these days) doesn't actually compress the audio at all. Instead it converts the fax to a tagged image file format (TIFF) image and sends it, along with its own description information. The transmission is less likely to fail because the data can be re-sent if it gets lost. There may be provisions to duplicate the information if you want to add a second layer of protection to the transmission.

The only shortcoming of this codec is that it's pretty new. Not all carriers are updated with it yet, so you may have to wait until it is provided. Until then, you can always mix and match your codecs, sending faxes over G.711 and using G.729 for voice calls only.

Ordering VoIP Service

When you order VoIP service for a business, you need to meet certain hardware requirements, and the testing requirements for your VoIP system are similar to a TDM voice installation. The residential VoIP offerings, on the other hand, are designed to be as simple as possible, and they include plug-and-play hardware and software installations. The basic hardware and service requirements for all VoIP applications (whether for business or home use) are the same:

- You must have a dedicated Internet connection.
- You must have a VoIP phone system, or an adapter that can translate the VoIP protocol to a protocol that your existing phone system can use.

Larger applications for big businesses with hundreds of employees may include data integration, videoconferencing, and Web conferencing — all on the same Internet line. Of course, having these additional perks requires more Internet bandwidth and more complex hardware to manage all the individual pieces of the network.

If you want to use VoIP technology on a small scale (for home use), you may need to purchase an inexpensive VoIP adapter (it is actually a small VoIP-to-TDM gateway) and a DSL or digital cable Internet connection. You can spend as much, or as little, money as you want on your VoIP hardware.

The key element to keep in mind is that generally the more complex your hardware, the more features available to you. But also remember that with great power comes great responsibility, and more advanced hardware also places a greater burden on you for troubleshooting any issue you may encounter.

Choosing your VoIP hardware

The hardware you need to buy for VoIP service depends on the complexity of the VoIP network you create. If you have one phone, you need a single piece

Capacity concerns are based on concurrent calls

VoIP is a very fluid and unstructured protocol when it comes to bandwidth consumption. When speaking about everyday TDM voice circuits, your biggest concern has to do with how many minutes you plan to talk. With VoIP, the question is more frequently about the number of concurrent calls your system can accept.

The TDM world is very structured; no matter what you do, a normal voice T-1 line can only handle a maximum of 24 phone calls at any one time. VoIP is a completely different animal, and depending on your codec, you may be limited to a few as 15, or as many as 76 calls on the same T-1 of bandwidth. Your carrier will assign ports based on your needs. Telling your VoIP carrier that you want to talk 100,000 minutes a month (as you would with a TDM order) means nothing for provisioning, but it may influence the per-minute rate the carrier offers you. With VoIP it's conceivable (if unlikely in most situations) that you could make 100,000 minutes of calls within a time frame of 2 hours a day, over two T-1s using G.729.

of hardware. More likely, however, you want a complex network of voicemail servers, additional end users, and a host of advanced features. If so, you need a small squad of hardware to take care of all your requirements.

Speak to your hardware vendor for assistance on setting up these more complex systems, or for a referral to a business that specializes in VoIP technology and hardware if your vendor isn't familiar with the technology.

Here's a list of hardware you can expect to purchase:

- **An Analog Telephone Adapter (ATA):** An ATA is the most basic level of hardware required for home applications and is installed between the wall jack for your dedicated Internet connection and your regular telephone. ATAs convert the VoIP signal coming from your Internet connection into a signal your phone can use. After an ATA is in place, you dial out and receive calls just as you did before you had VoIP. An ATA is a common piece of hardware for home use; you will probably need something more complex for a business application with more features.
- **A softphone:** A softphone is a software program for your computer that enables it to function like a phone. Softphones usually look like the keypad for a cellphone and use the speakers and a microphone on your computer to transmit and receive your conversation. I have used a few of the softphones available, and I don't think the quality is wonderful. Usually, the biggest limitation to quality is the fact that if the speakers and microphone for your computer are crummy, so is your VoIP conversation. If you have a great computer and are comfortable with using your computer to make calls, a softphone may be your phone of choice.
- **A gateway:** A gateway is a device that converts VoIP to TDM and TDM to VoIP. If you are deploying VoIP service and have any non-VoIP phones that need to use the service, you need a gateway.
- **An Edge Proxy Server (EPS):** An EPS is the first piece of hardware a VoIP call comes into contact with on your network; it's used for larger VoIP applications. This server receives and routes the signaling and may also route the RTP. Because the RTP and signaling are separate, the RTP could be directed to another piece of hardware for voicemail, music on hold, or handling by an *automatic call distribution (ACD)* program that prompts you to "Press 1 for Sales, press 2 for Customer Service . . .", and so on.
- **A media server:** Media servers are computer servers that receive and process the RTP stream in a VoIP call. The media server may be a part of the EPS, or it may be a completely separate device.

Conserving bandwidth by controlling the RTP stream

The largest consumption of bandwidth on a VoIP call is the RTP stream, because it has to house and transport 64 kbps of voice communication. The signaling is very small in comparison and is the only real part of a VoIP call that must be handled by the EPS. Figure 15-4 shows the bandwidth requirements for a VoIP call where the RTP is transmitted through the EPS.

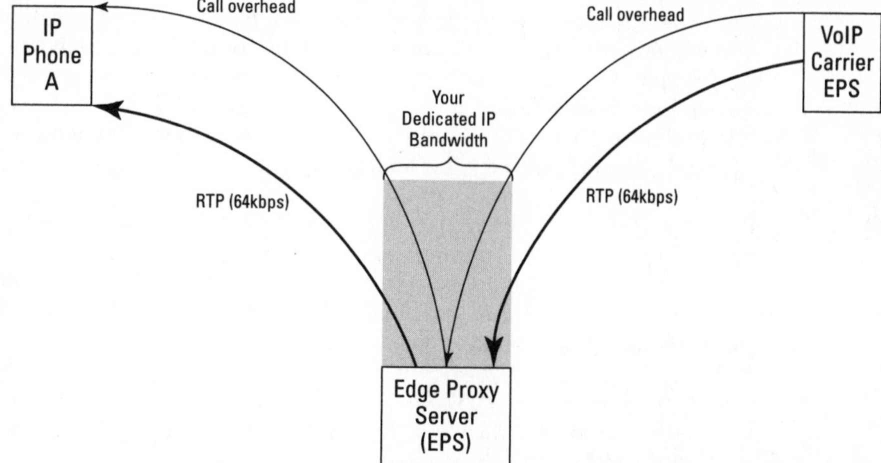

Figure 15-4: Bandwidth consumed by the EPS during a VoIP call.

As you can see, both the 64 kbps RTP stream and the smaller overhead stream are handled by your EPS. If the IP phone you are transferring the call to is not part of your local area network (LAN), the call is sent back to your IP connection, and the RTP alone ends up using at least 128 kbps — for just one call.

Getting more information on VoIP protocols

I understand that VoIP isn't for the faint of heart. There's a lot to know before you make a commitment to tying your phone service to your Internet connection. Perhaps because VoIP is so tied to the Internet, I thought it might be good to give you some Internet-based resources to answer your questions. In addition to picking up a copy of *VoIP For Dummies*, by Timothy V. Kelly (Wiley), you can also find more information online:

- A good Web site for information about VoIP protocols, H.323, and SIP is `www2.rad.com/networks/2001/voip/prtcls.htm`.
- If you're looking for a good Web site with information to answer your RTP-related questions, visit `www.cs.columbia.edu/~hgs/rtp/faq.html`.

Passing the Interoperability test

After you order an Internet connection and purchase the required hardware, you may still have to pass an *interoperability (InterOp) test.* The only way you should be allowed to skip the InterOp test is if you are ordering very basic, preconfigured plug-and-play hardware from your carrier. If your carrier sends you the hardware, you should receive instructions on setting it up. In such a case, InterOp testing isn't as crucial.

However, the InterOp test becomes very important when you have ordered your own hardware and you are beginning a new VoIP relationship with a carrier. VoIP carriers have their own default transmission configurations, and you will either be able to modify that default, or be forced to live with it, depending on the construction of your carrier's network. Your carrier's switch may have a global setting for features like *SIP-T;* SIP-T sends ANI Infodigits, which identify whether a call was made from a pay phone, hospital, or a prison (to name a few locations).

Before you can enter InterOp testing, you have to fill out an InterOp questionnaire. This document, supplied by your carrier, not only tells your carrier important information about your network configuration and projected call volume, but it also shows your level of knowledge of VoIP. A standard InterOp questionnaire asks you for some of the following information:

- **Your IP protocol (H.323 or SIP):** Your carrier needs to know this info to establish the most basic level of connectivity between your router and your carrier.
- **The H.323 or SIP hardware, location, and IP address:** The physical location of your hardware may determine which Edge Proxy Server your carrier uses. Most carriers don't have thousands of servers to receive customers across North America, so the location of your hardware determines which regional server is assigned.
- **RTP media gateway:** If you're using a separate media gateway to transmit your RTP stream, your carrier needs to know where it is located, its make and model, as well as its IP address.
- **Other SDP and/or RTP hardware:** Many customers have a separate server with Automatic Call Distribution features so you can "Press 1 for Sales," or "Press 2 for Customer Service." If you're planning to use different pieces of hardware to provide additional features, your carrier needs to know the function of each piece, as well as info about the make, model, and IP address. The information you provide about your network design helps the carrier to troubleshoot complex issues you may encounter later.

- **Codecs you want to use:** Your carrier may not offer every codec you want to use, so be sure to ask your carrier which codecs are available. For example, the T.38 FoIP codec exists, but hasn't been fully deployed by all carriers.

The InterOp questionnaire has even more questions on it, but its main purpose is to communicate to your carrier your business model and the hardware you are using to provide VoIP service. Many carriers also ask for a diagram of your network and the hardware involved.

InterOp testing for compatibility

VoIP has millions of options, variations, and nuances that can prevent you from completing a successful call. The InterOp questionnaire allows your carrier to determine all the types of calls you will be making. The point is to test every variety of call your hardware is capable of, preventing protracted troubleshooting in the long run. Your carrier determines the criteria for your InterOp testing and what decides when the test is completed.

This process can take as little as 30 minutes, or as long as months. The more knowledgeable you are about your hardware, the less time consuming InterOp testing is.

If you aren't required to complete an InterOp test and are using hardware that wasn't provided and configured by your carrier, you should ask for an InterOp test. Just because your carrier thinks it isn't necessary doesn't mean that you can't require it. If the carrier doesn't require an InterOp test, call the carrier's customer service department and open a technician assistance *(tech assist)* ticket to schedule some testing time with a VoIP tech.

If you are new to VoIP, hire a qualified technician. You may be having a wonderful time discovering the nuances of SIP, but eventually you will reach the limits of your knowledge and begin grasping at straws. Simply pay someone else with the experience to finish the job so that you can eliminate your frustration with VoIP.

Hearing VoIP Quality Issues

Like any other type of phone call, VoIP calls can experience problems, but the two concerns that are more prevalent within VoIP are linked to the very structure of VoIP. A plain-vanilla TDM (non-VoIP) call transmits sound on a circuit, with continuous, unimpeded information being sent in both directions. A VoIP call, however, transmits sound in small packets over an IP network that also has millions of other packets within it that are traveling in every direction.

The very idea of a transmission network filled with VoIP packets that come and go, finally arriving at the far end without losing any information, and in the correct sequence, seems like an achievement. But this is exactly what is expected of VoIP transmissions. As long as your LAN, your carrier's network, and the Internet aren't congested, everything should work fine. If any of these three elements do have congestion, your transmissions could arrive late, out of sequence and minus a few packets. The following sections go into what to do.

Getting beyond latency

Latency is a delay in the transmission of your data. Even though voice data is converted into digital packets, in order to be understood, it must come out the other end as a continuous transmission of sound. If a second of the conversation drops because it arrived late or was lost along the way, your call quality will quickly erode to static or sections of speech will be eliminated (a call treatment called *clipping*).

Latency can be caused in the coding and decoding process of your RTP stream, as well as the slowing down of a packet because of the geographic distance it is being sent. Latency issues are generally intermittent and can be caused by a variety of issues. In fact, a consistent latency problem is uncommon; if you are so lucky, you can adjust your hardware to accommodate it.

Dealing with jitter

Jitter refers to the fluctuations in latency. VoIP packets may experience a 10 millisecond (ms) delay during off-peak times, but during peak traffic hours for your carrier, your Internet provider, or your LAN, the delay may raise up to 60 ms. Jitter could cause portions of your call to disappear. If you hear short, split-second pulses of dead air in your conversations, these gaps are places where packets of data were supposed to be.

This problem is even more noticeable if your VoIP service is integrated on your LAN that also carries data, video, and Internet traffic. Your transmissions may sound great at times, but when your co-worker decides to download *War and Peace,* and your boss is videoconferencing with your New York office, you will probably see an instant jump in latency.

If your network tries to accommodate for this fluctuation, it has to readjust when the last page of *War and Peace* finishes downloading, and then again when your boss's videoconference ends.

A *jitter buffer* is a function within your VoIP server that collects data packets, prioritizes them, and sends them out in sequence. They can help reduce jitter in VoIP transmissions, but excessive congestion will overload the buffer and result in lost packets.

If you're integrating VoIP into your existing LAN, hire an experienced professional to validate your design. The assignment of one piece of hardware in the wrong place will cause you prolonged pain and suffering.

Moving your existing telephone number to VoIP

Local number portability is the government-sanctioned freedom you have to move your phone number from one local exchange provider (like Bell South, Verizon, or Qwest, for example) to another. In order to use VoIP on your old phone number, your new VoIP carrier must have ownership of it. All carriers in the U.S. hand off their calls to the local carrier that owns the dialed phone number. This means that until your VoIP carrier is in charge of your number, you won't receive any inbound calls on VoIP.

The process to migrate, or *port* your phone number from one carrier to another can be tedious, time consuming, and supremely frustrating. The requirement to handle the acquisition and release of local numbers is brand new for most carriers, so they're still building their processes and procedures.

The migration of a normal, noncellular phone number may take anywhere from 5 to 50 days, depending on the carriers you are porting to and from. If you are beginning to port your phone number to a VoIP provider, keep an open mind, a positive attitude, and follow up every week to see how it is going.

The current trend in the industry is to charge for the migration of telephone numbers. Don't be shocked if your carrier charges $20 or more per phone number for a migration. The evolving migration process can include many delays that must be resolved. Issue resolution requires manpower, and manpower increases overhead. That means you have to pay more. As VoIP develops, expect more fees and charges to crop up as carriers respond to the market and protect their profitability.

Part VI

The Part of Tens

The 5th Wave
By Rich Tennant
GAREN HELPS DEVELOP THE FIRST VOICE OVER TURNTABLE PROTOCOL
© RICHTENNANT

"Hello?! Hello, Phillip?! You're breaking up! Listen, put a penny on the tone arm and turn the speed up to 45 RPM!"

In this part . . .

The Part of Tens contains three chapters of tens (I guess that's one part of 30?), each with ten snippets you can use. Chapter 16 warms you up to telecom lingo, while Chapter 17 cools you down by giving you some realistic expectations of how the telecom industry actually muddles along. And Chapter 18 offers real-life bits of wisdom and helps you find the people, Web sites, and built-in safety nets that can make things work. In general, if you keep in mind all the information you discover in these three chapters you may just maintain some semblance of sanity in an industry that can be a little on the chaotic side. If all else fails, you can always turn to the Magic 8 Ball for consultation.

The appendix ends the book with all the information you need to build loopback plugs. These little guys are the best testing equipment available; all the better, you can build them for less than $10 and they're simple to create. If you have a dedicated circuit, you need to read this appendix; in just a couple of pages you can make your life, and the life of your carrier, at least 95 percent less stressful.

Chapter 16

Ten Acronyms and What They Really Mean

In This Chapter

- Knowing your LEC from your ANI
- Sneaking around with NASC
- Getting explanations with an RFO
- Finding your CSU and ordering your CFA
- Making magic with PFM

The world of telecom can seem like an exclusive club at times, mainly because its members throw around a lot of secret code words. If you don't use the right jargon, you are deemed an outsider and may not be treated with the level of respect you deserve.

The good news is that familiarity with just a handful of terms will get you into a comfortable middle-class position in the pecking order, where you can get both respect and help. The ten acronyms listed in this chapter are common to telecom technocrats, so if you use them, your carrier's technicians will think you're one of them.

Getting to Know Your LEC

The term LEC (pronounced *lek*) stands for local exchange carrier. LECs are local phone carriers, such as Verizon, Bell South, Qwest, or Southern Bell Company (SBC). If you are speaking to your long-distance carrier, you can refer to the company that provides dial tone and phone numbers for you as a LEC.

When you call your long-distance carrier to chat about a new phone line that needs long-distance service, simply say, "I just got a new phone number from my LEC that you need to set up on your network."

Understanding ANIs

You can either pronounce ANI as *ay-en-eye* or *ann-ee,* but in the end, your ANI is a phone number. During trouble reporting, if a technician asks for the origination and termination ANI, you just need to provide the phone number that originated the call and the phone number that was dialed (or where the call was intended to terminate).

This term can be used very often. When you call your LEC, you could ask, "Are all of my ANIs set up with AT&T for long distance?" If you have many phone lines and a phone system that pulls them all together, you could use the term when opening a trouble ticket with your long-distance carrier. Say, "We have about 20 ANIs going through our phone system, so I don't know which line actually originated the call. But it was made to the *term ANI* at exactly 3:27 p.m. today." (Term ANI stands for terminating ANI or terminating phone number.)

Getting Firm with an FOC

The term FOC stands for *Firm Order Commitment.* You can pronounce it by its individual letters *(eff-oh-see)* or call it a *fawk*. The FOC is the document that tells you the date on which your order will be completed. This goes for all orders in telecom, from the installation of huge dedicated circuits to the conversion of a single phone line to a new long-distance carrier.

Say, "Our order has been pending for a few days, do we have a FOC yet?" The same question is sometimes asked, "Did we get FOCed on the new T-1s yet?" (This sentence is generally spoken using the second pronunciation.)

NASCing Your Numbers

A NASC (pronounced *nask*) is more than just an administrative group in telecom, it's also a threat. NASCing is the act of using the *Number Administration and Service Center* to forcibly extract a toll-free number from a carrier. If you're in the process of moving all your toll-free numbers to a new carrier, and are met with resistance from your old carrier, you can eliminate the childish bickering and rejections and just NASC your numbers. The luxury of NASCing costs about $40 per toll-free number.

After having your toll-free number migration rejected numerous times for silly reasons, call your new carrier and say, "I am tired of this game of patty-cake! Just NASC the numbers and be done with it." As long as your old carrier doesn't want to NASC the numbers back and begin a game of tug of war, NASCing is your quickest solution to toll-free migration problems.

Getting an RFO

As in most industries, a problem found generally turns into a cover-up begun. An RFO, Reason for Outage, is a document that provides the official Word on the cause of a network issue that prevents your service from working. If your carrier's switch sustains a direct lightening strike, preventing you from using your phone service for three days, it is acceptable to ask for an RFO.

An RFO is a document that your carrier (rightfully) thinks may be used against it in a court of law. Because of this, you have a very good chance of receiving a document that is so watered down and vague that it has no real substance to it. Any RFO you receive will only tell you what you already know.

Sound like a pro when you say to a representative at a carrier, "My boss is steamed that our T-1 was down for five days; can you get me some sort of an RFO so that he knows I was pushing to get the issue addressed and repaired?"

Getting Your Hands on a CSU

The *Channel Service Unit* (CSU, pronounced *see-ess-you*) is a little device about the size of a cigar box that's used on dedicated circuits. It is the first piece of hardware you need to make your dedicated circuit work, and its daily job is boosting the signal coming in from your carrier. The CSU is also the first piece of hardware used to validate the continuity on a dedicated circuit.

If you want to impress a technician without sounding obvious, say, "Wow, you can test to my CSU without any problem? That's great! I'll call my hardware vendor right now to fix my phone system and then cut out for lunch."

Making Sure You Get a CFA

CFA (pronounced *see-eff-ay*) stands for Carrier Facilities Agreement, and the term is associated with ordering a dedicated circuit. It refers to the specific location into which your circuit cable will be wired. The CFA actually lists the physical location of your carrier's point of presence (POP), as well as the bay, panel, rack, and slot to which your circuit is assigned. It's a brief, one-page document, but it is very specific. The CFA is also one of the first benchmarks in the process of setting up dedicated service. When you receive the CFA, you can rest assured that you're halfway to completion.

When you're checking on the progress of your dedicated service, the next time you talk to someone at your carrier, say, "It's been three weeks since we sent you the order for our new circuit in Dallas. Do we have a CFA yet?"

Being a Part of the PFM

Pure Friggin' Magic refers to the mysterious events that resolve a problem, sometimes a huge outage, with no apparent action being taken by anyone. It is as if someone's fairy godmother came down and sprinkled pixie dust on the network and suddenly it all returned to normal.

PFM may indicate that someone secretly repaired an issue that he or she caused to begin with. This is actually more desirable than the other possible cause of PFM — an intermittent issue that will come back again soon. If the problem crops up again, deem it as an intermittent issue and troubleshoot it vigorously.

Tell your boss the following, "I don't know what happened! Our carrier and our hardware vendor swear they didn't touch a thing to fix our problem. We seem to have had a case of PFM."

If you work for a boss or organization that eschews swear words, don't use this expression.

Getting Your CICs

The CIC (pronounced *kik*) is the four-digit number that your local carrier uses to identify your long distance network. This number is also part of the code you can dial if you want to use another long-distance carrier to make a call. For example, if you have MCI for long distance but want to use AT&T the next time you call London (maybe you heard that AT&T has a great network to the U.K.), you simply dial 10-10-288, plus the international phone number to complete your call.

If you do make a call over AT&T to London, and the number you are dialing from isn't set up with AT&T, you may be charged a very extreme per-minute rate for that call. Don't use another carrier's CIC unless you are willing to pay up to $5 per minute for the luxury.

When speaking to your local carrier to change the long -distance on your phone lines, say, "Please set up all of my lines to CIC 0333. I just signed up with Sprint and I don't want to wait for the time it takes them to send the order to you."

Chapter 17

Ten Troublesome Telecom Traits to Avoid

In This Chapter

- Avoiding finger-pointing
- Dealing with outages and credits
- Dealing with fraud
- Living with projects
- Dealing with dementia
- Calculating the cost of invoice inaccuracy
- Threatening to sue

Every industry has its little list of pitfalls, and here are ten of the largest ones that you may find yourself dealing with during your adventures in telecom. They are all dangerous, painful, and surprisingly easy to find, so keep your eyes open and try to avoid them.

Finger-Pointing Your Way into a Corner

You don't like to be told you're wrong. Heck, no one does — especially carriers and hardware vendors. There are many variables in telecom, and although you may think you've isolated a problem and can pinpoint its source, you may want to avoid doing so. There is always a quirky possibility that a problem you find in one location may be the result of something wrong in a seemingly unrelated part of the network.

If you sound too sure that you know what a problem is, and who is to blame, a tense situation can quickly deteriorate into a series of verbal attacks and pointed words. If your troubleshooting team is at odds with each other, you are doomed to either failure or a very dicey bit of diplomacy that makes the Mid-East peace talks look like a cakewalk.

If emotions run high, end the troubleshooting, separate the people that are fighting and adopt a diplomatic stance.

Expecting a Credit After an Outage

It has become very fashionable lately to believe that every inconvenience a person suffers should be matched with an appropriate amount of monetary compensation. This assumption, just like the societal acceptance of notched-lapelled tuxedoes, is patently wrong.

Most carriers have a clause in their service contracts that prevent customers from requesting credits for outages suffered while using their service. Think about it from the carriers' standpoint; they have thousands of customers without service during an outage, and the financial impact on carriers is probably ten times the financial loss you are experiencing.

If you do receive a credit, be thankful. Understand that the credit is a courtesy. The carrier isn't recovering the money from someone else; it's losing money to keep you happy.

Ignoring the Facts: Fraud Is Not Free

If you are the victim of fraud on your phone system, you aren't guaranteed credit from your carrier. Legally, every call placed over your phone line is legitimately your responsibility, and your carrier has every right to pursue payment, even if no one at your company made the calls.

Some carriers are confident of their fraud departments and state that they will credit the accounts of any customers that experience fraud, but that is the exception and not the rule. Protect your phone system and watch your usage, because in the end, you are still financially responsible for any fraud that is generated from your phone lines.

Not Accepting Admitting Defeat When an Order Turns into a Project

If anyone refers to your order as a *project,* you're in trouble. The word is an omen that signifies delays and complexities.

If you are ordering five dedicated circuits and your carrier tells you that it is being handled as a project, you should just tack on another 5 to 15 days to your timeline without blinking. Adopting a Zen-like sense of peace and acceptance is a good move at this point.

There are generally signs that identify when an order is treated as a project, like the quantity of circuits ordered. Trying to break an order of five circuits down into an order of three circuits and another order of two circuits won't help you out. If all the circuits are going to the same building, the carrier will aggregate them and still treat the order as a (dreaded) project.

Having Expectations That Go Beyond Reality

It happens every day. You tell a member of a carrier's sales staff that you need a dedicated circuit installed in the back hills of Alabama, with an interactive Web site allowing you to view the total calls being processed, the average call duration, connection time, and the ability to reprogram your circuit to hunt in a different sequence depending on your call volume. You need this perfect system in five days.

You hear in response, "Of course! No problem!" Salespeople are out of touch with reality. The sad part about this person's dementia is the fact that the salesperson is trying to take you along with for the ride. You will abruptly arrive at the off ramp to reality 10 days after you sign the contract, when you are told by the provisioner that the carrier doesn't offer interactive Web sites, that the projected installation date is in 45 to 60 days, and that the reporting you want isn't available.

If something sounds too good to be true, it probably is.

Expecting Mother Theresa

There are no Calcutta nuns that I know of currently working in telecom in the U.S. There's also nobody like Special Agent Fox Mulder running around, working 24 hours a day to unravel the mystery of why you can't call your Aunt Maybell in Des Moines.

The people who work at phone companies are just like you. They put in their eight hours and gladly leave on time. You shouldn't expect everyone you speak to at your carrier to have an emotional connection and conviction to resolve your issues. If you do have a pleasant interaction with someone who is sharp and attentive, be sure to thank this wonderful person profusely while you quietly write down his or her direct phone number.

Great people can turn a three-day nightmare into a five-minute annoyance. Send them a *you're so great* letter, and they will probably make themselves available for you when you need them later.

Not Paying Attention to Smaller Companies

If you want a stable relationship with a carrier, you may gravitate toward well-established companies like AT&T, MCI, or Sprint. If service is what really matters to you, you might need to refine your criteria.

Large telecom companies that have been around forever are often disjointed amalgamations of smaller companies they consumed, merged with, split off for special projects, or inherited from companies they consumed, merged with, or were split off for special projects. To avoid the incongruous workings of the main company, find a smaller, more personal connection with a smaller carrier. You could get lucky and end up with a less antiseptic version of customer service.

Either contract with a *switchless reseller* that uses one of these carriers to handle its traffic or do some research to find any special division within the big carrier that has been designed to interact with you on a human level. Either way, you might just get the best of both worlds.

Forgetting to Do the Math

Generating a correct invoice is a monumental task for every carrier. The rates could be wrong, the monthly charges could be double billed, or the carrier could be billing you for features you never ordered. That list just covers *some* of the aspects your carriers have control over. Taxes are another issue and generally can't be explained without some shades of gray, even by the government organizations that administer them. Check your phone bill every month against your contract to ensure that every rate and charge is correct. The likelihood is high that over the course of a year, you will find at least one thing that is wrong.

Falling for the Standard Interval Shield

You can always tell when technicians and provisioners are getting anxious, because they begin to find ways to protect themselves. For example, if a trouble ticket should reasonably be resolved by now, or if a project isn't moving forward, a technician might reference the *standard interval* assigned for the project.

Here's an example: You call your carrier because you want to change the ring-to number on your toll-free number. You may hear, in response to your request, "Our standard interval is 48 hours, so don't ask me again until then." The task actually takes about five minutes to accomplish as soon as someone begins working on it, but the carrier is probably too overloaded to get started on the project.

If you receive the *standard interval* line from your carrier, try to ask (in a nonthreatening and lighthearted way) why your order is being shot down. You could phrase your question like this, "What kinds of fires are you dealing with over there?" or "What are the other things on your plate?"

If you ask the right questions in the right way, you can get a better sense of the pressure cooker in which your technician is currently working. Your carrier may have just laid off five people in the department and three more may have left for better jobs at Krispy Kreme, so the people left are hopelessly swamped.

Understanding the environment of the people who actually do the work at your carrier will make your life easier. They will be more comfortable telling you the truth rather than the official corporate-authorized story, and you can then adjust your expectations accordingly so everyone doesn't feel crushed. So the next time you hear someone give you the *standard interval* routine, recognize it as the cry for help that it is.

Demanding to Sue or Take Legal Action

Nobody in telecom wants to receive ten faxes from midnight to 5 a.m. every day from some spammer selling Viagra, but making it your mission to hunt the perpetrator down and take legal action is counterproductive.

Resolving a harassment call (like the repetitive fax call) can take as little as about 24 hours if you simply report the offender to your local phone carrier and let someone else do the hard work. Carriers can find the call, track it back to the long-distance carrier that originated it, and have the long-distance carrier force its customer to stop bugging you.

You can prevent some of these calls from dialing your home by listing your phone number in the National Do Not Call Registry. This is only an option for residential phone numbers. You can find out more about it at `www.donotcall.gov/default.aspx`.

Your timeline is completely different if you decide to sue the person calling you. It's a long and painful process that may or may not yield results. Rather than relying on the honor of the carriers to help you (which isn't hard to work with) you are investing yourself in a legal system that must allow for due process and could take months to accomplish the same thing.

Chapter 18

Ten Places to Go for Hints and Help

In This Chapter

- Talking to your long-distance carrier or salesperson
- Discussing things with your hardware vendor
- Getting local support
- Visiting the Magic 8 Ball oracle
- Heading to the manufacturer
- Searching the Internet
- Escalating through the ranks
- Taking it to the competition
- Going back to where you started

A time will come when you will need more information than you currently have. Troubleshooting an intermittent problem, validating your rates, or investigating new technology to ensure that it's a fit for your company can all exhaust your telecom knowledge.

This chapter covers ten sources of information that can help you overcome when you are working above your telecom comfort level. Use them, accept them, and with perseverance, you can overcome any challenge.

Calling Your Long-Distance Carrier

If you are trying to find a new solution to your telecom needs, or are working through a potential trouble issue, your long-distance carrier is a good place to start. The only time a representative from your carrier isn't a good resource for you is when the information you seek relates strictly to your local service (if you want information about directory listings or caller ID, talk to your local carrier).

Long-distance carriers can even handle questions that relate to your phone system. The carrier may be able to schedule some testing time by opening a *tech-assist trouble ticket.* The tech-assist trouble ticket enables the long-distance carrier to work directly with your hardware vendor to resolve a problem.

Smooth Talking with Your Telecom Salesperson

Your telecom salesperson is helpful for both general technology questions and trouble issues. This person may not have all the technical information, but nevertheless the salesperson is often the most helpful person you know.

That's right. It's the job of salespeople to make the seemingly impossible happen. They have tough quotas to fill, and when you have to make things happen, you get to know the names of the people who can help. So talk to your salesperson for contacts and logistical information on getting something done.

Your sales contact will direct your questions on new technology to a presales technical engineer, and your trouble issues can be escalated to a customer service manager. The salesperson has massive incentive to help you, because he or she gets paid based on what you buy. Even if the service you are interested in won't work for you, by tracking down information the salesperson is hoping to retain you as a customer.

Talking to Your Hardware Vendor

Your hardware vendor has an incentive to tell you all about the latest new technology. If, through the course of explaining a glitch with your frame relay data transfer service, the vendor rep can convert you to MPLS (see Chapter 14), the vendor can get what he or she wants, too — to sell you hardware for the transition.

Vendors tend to have more hands-on experience with new technologies and can give you an unbiased assessment of what works and fails in the world of telecom. Troubleshooting should *always* include your hardware vendor if your carrier suspects your hardware is involved in the problem.

Visiting the Local Calling Guide Industry Web Site

The industry Web site, Local Calling Guide (www.localcallingguide.com), provides you with the identity of the local carrier, OCN number, and LATA by simply inputting the area code and the first three digits of any phone number in the U.S. This simple task can help to find your local carrier, or the local carrier of someone you are trying to call. The site also enables you to see a list of area codes by local carrier or by LATA. You will actually be surprised how helpful this Web site is.

Using the Magic 8 Ball

You may know the Magic 8 Ball as a small toy made by Tyco Toys (a subsidiary of Mattel, Inc.), but to the universe of telecom, the Magic 8 Ball is the oracle of the greatest possible likelihood. The little 20-sided die inside the blue inky fluid may have a 50 percent representation of positive outcomes, but it still is an enjoyable and quick answer for any question you have. Question: Will my circuit be installed by Christmas? Answer: Outcome unclear.

Going to Manufacturer Web Sites

The vast majority of simple questions you have about your hardware can be answered by visiting the manufacturer's Web site. You can also find out about hardware upgrades and options for the specific model of hardware you have, expansion capabilities, and more.

Many Web sites provide a visual schematic to show you how everything in your phone system connects together. If you are familiar with a manufacturer's Web site, you can alert a carrier, who may be trying to troubleshoot the hardware but is unfamiliar with the setup.

Searching the Internet

The Internet is more than just a place to Google your own name, buy airline tickets, or do your Christmas shopping; it can also provide you with much needed information. There are numerous Web sites that discuss specific telecom hardware, technologies, and the problems related to each of them. Try searching for a keyword that is specific to the technology you are using. For example, if you are looking for information about VoIP, search for the specific codec that you are having difficulty with instead of just *VoIP transmission problems.*

Check out Chapter 16 for a quick peak at some of the most common acronyms in telecom. Here's a quick search technique for finding telecom definitions on the fly. Visit Google and type **define:*keyword*** in the search box. For example, if you want to know what LATA means, type **define:LATA**. Try this technique the next time you're on the phone with a technician and he or she says something you've never heard of before.

Using Your Escalation List

You should have an *escalation list* for all of your carriers for both provisioning and trouble reporting issues (these are generally different departments within carriers and have unique escalation procedures). The escalation list is a list of names and numbers. You start with the lowest person on the totem pole and work your way through the list until you talk to someone who can help you manage your problem.

Carriers have a hierarchy of technicians that receive trouble issues as they are escalated. I would like to say that this is the same with orders, but, unfortunately, the opposite is true. Escalating up the chain of command in provisioning will generally connect you with people that have more responsibility, but less knowledge of how to actually get things done. In the end, they will simply walk down the hall to the person you chatted with first. Of course, sometimes raising the visibility of a problem is all you need to do to get it worked at your desired level of urgency.

Taking Your Questions to Another Hardware Vendor

If you're looking for guidance on a new technology, or troubleshooting a piece of hardware that your company inherited before you found a vendor, the system in question may not be the strong suit for your current hardware vendor. In this case, you need to find a new hardware vendor. The world of telecom vendors isn't that immense, and so your vendor may even refer you to another vendor for information about your specific system.

If your hardware vendor doesn't know anyone or is less than inclined to give you a referral, contact the manufacturer directly for a list of authorized sales and service companies. ***Hint:*** You can usually find such lists on manufacturer Web sites.

Starting Over

Problems and inquiries grow like insulating foam, taking up space wherever there is the less resistance. Checking all variables is wonderful, but if you've ever lost your car keys, you know that the first place you looked (if stomping your feet and throwing stuff around in a frenzy because you're late constitutes "looking") is usually where you end up finding the keys.

The point is that you can easily overlook something when you're in a panic that is clear as day when you've calmed down a bit. People commonly troubleshoot a problem for days on end, only to find out that there was something as simple as a configuration issue to blame.

When you are at your wit's end, stop, bring all the key players together for a conference call, and slowly go over all the information about your service. Start with how the service is configured, move on to a discussion of the hardware you are using, and move on from there by describing the specific issue and sharing the most detailed information you have about the variables you believe could be affecting the issue. Maybe it'll be news to someone that you changed your configuration from ISDN to loopstart last month — and that breakthrough could be the spark that solves the issue.

This same approach also works when you are reviewing new technology. Don't try to make your application fit the new technology. Instead, objectively see whether the technology can successfully fulfill your application needs. You may think that some complex technology like MPLS or VoIP sounds very high speed and sexy, but if neither option solves your problem, it is just a burden.

Appendix

Making a Loopback Plug

In This Appendix

- Introducing male loopback plugs
- Setting up female loopback plugs

A loopback plug is the cheapest piece of test equipment you can own. It consists of a small plastic connector and about 10 inches of wire. The loopback plug assists in insolating problems within your circuit and provides a device for your carrier to run to during intrusive testing. This appendix gives you all the information you need in order to get a functioning loopback plug. Check out Chapter 13 for more information on using a loopback plug to facilitate troubleshooting.

Making a Male Loopback Plug

The male loopback plug for a T-1 circuit consists of an RJ-45 connector with pin 1 connected to pin 4 and pin 2 connected to pin 5. You can easily make your own male loopback plug — or have your hardware vendor make one for you.

The next time a technician from your hardware vendor is in your office, have the technician make a loopback plug and leave it hanging on a nail in your phone room. If the tech is feeling generous, he or she may even make both male and female loopback plugs.

Here's what you need to make your own male loopback plug:

- **RJ-45 connectors:** You can buy packs of RJ-45 connectors from any electronics, hardware, or home improvement store for anywhere from $5 to $10.

- **Crimping device:** This tool enables you to secure the wires into the RJ-45 connector and costs between $20 and $55. Nicer crimping tools also allow you to work on RJ-11 connectors, which are used for standard telephones, fax machines, and analog modem lines.
- **Wires:** You use wires to input into the connector. You can cut them from any flat phone or modem cable. After you cut a 6-inch section of cable and remove 2 wires, you need to insert them into the RJ-45 connector (see Figure A-1).

This configuration allows you to connect the transmit wires 1 and 2 to the receive wires 4 and 5. After inserting the wires into the slots, simply crimp down the copper blades in the connector to secure the wires and complete the connection. You now have a male loopback connector.

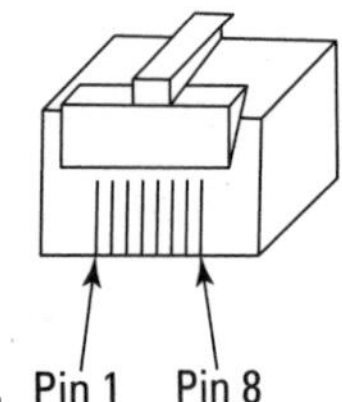

Figure A-1: The pin configuration for a male RJ-45 plug.

Making a Female Loopback Plug

To build a female loopback plug, you need to connect the terminal for pin 1 to pin 4 and pin 2 to pin 5. Here's what else you need to make a female loopback plug for a T-1 level circuit:

- **A standard category 5e (Cat 5e) jack:** You can find Cat 5e jacks in the same general area as the RJ-45 connectors at your local home improvement or electronics store. Some Cat 5e jacks even come with their own push-stick to secure the wiring. The pin locations are listed on the side of the jack. After pushing the wires into place, simply cut off the excess and close the jack.
- **Crimping device:** Refer to the previous section for more information about crimping devices.
- **Wires:** Of course you need wires.

Insert the RJ-45 connector into the CAT 5e jack to keep both of your valuable pieces of test equipment together. If you leave enough extra wire on the RJ-45 loopback, you can use the loop on the wires to hold the plug on a hook, screw, or nail in your phone room. This will ensure you have both connectors available for use and at the ready for whenever you need them.

Finding online resources

Looking for more information about setting up a loopback plug? Check out the following online resources:

www.nullmodem.com/RJ-45.htm

www.cisco.com/warp/public/471/hard_loopback.html#lb_png

The loopback devices listed here are specifically used to test T-1 level voice circuits. The same RJ-45 connector and Cat 5 jack can be used to test other types of circuits, but the pin configurations are different:

- **Loopback wiring for a 56K CSU/DSU:** Wire to connect position 1 to position 7, and with the second wire, connect position 2 to position 8.
- **Loopback wiring for a FastE circuit:** Wire to connect position 1 to position 3, and with the second wire, connect position 2 to position 6.

Index

• Numerics •

• A •

• B •

• C •

• D •

• E •

• F •

• G •

• H •

• I •

• J •

• K •

• L •

• M •

• N •

• O •

• P •

• R •

• S •

• T •

• U •

• V •

• W •

• Z •

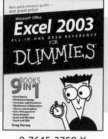

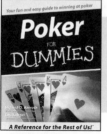

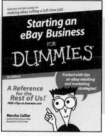

SPORTS, FITNESS, PARENTING, RELIGION & SPIRITUALITY

0-7645-5146-9

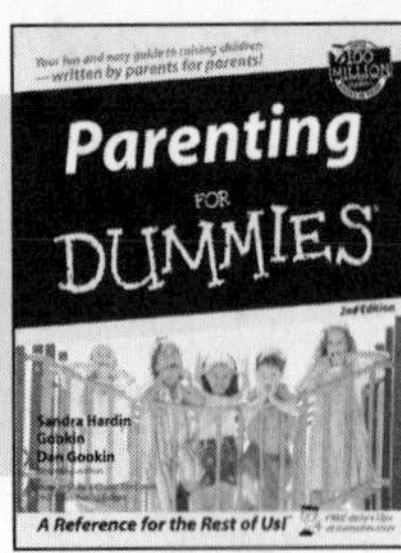

0-7645-5418-2

Also available:

- Adoption For Dummies 0-7645-5488-3
- Basketball For Dummies 0-7645-5248-1
- The Bible For Dummies 0-7645-5296-1
- Buddhism For Dummies 0-7645-5359-3
- Catholicism For Dummies 0-7645-5391-7
- Hockey For Dummies 0-7645-5228-7
- Judaism For Dummies 0-7645-5299-6
- Martial Arts For Dummies 0-7645-5358-5
- Pilates For Dummies 0-7645-5397-6
- Religion For Dummies 0-7645-5264-3
- Teaching Kids to Read For Dummies 0-7645-4043-2
- Weight Training For Dummies 0-7645-5168-X
- Yoga For Dummies 0-7645-5117-5

TRAVEL

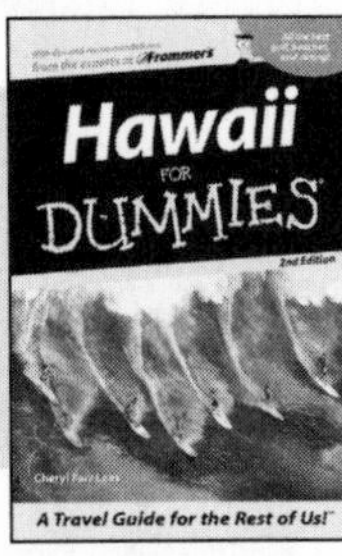

0-7645-5438-7

Italy For Dummies, 2nd Edition

0-7645-5453-0

Also available:

- Alaska For Dummies 0-7645-1761-9
- Arizona For Dummies 0-7645-6938-4
- Cancún and the Yucatán For Dummies 0-7645-2437-2
- Cruise Vacations For Dummies 0-7645-6941-4
- Europe For Dummies 0-7645-5456-5
- Ireland For Dummies 0-7645-5455-7
- Las Vegas For Dummies 0-7645-5448-4
- London For Dummies 0-7645-4277-X
- New York City For Dummies 0-7645-6945-7
- Paris For Dummies 0-7645-5494-8
- RV Vacations For Dummies 0-7645-5443-3
- Walt Disney World & Orlando For Dummies 0-7645-6943-0

GRAPHICS, DESIGN & WEB DEVELOPMENT

0-7645-4345-8

0-7645-5589-8

Also available:

- Adobe Acrobat 6 PDF For Dummies 0-7645-3760-1
- Building a Web Site For Dummies 0-7645-7144-3
- Dreamweaver MX 2004 For Dummies 0-7645-4342-3
- FrontPage 2003 For Dummies 0-7645-3882-9
- HTML 4 For Dummies 0-7645-1995-6
- Illustrator cs For Dummies 0-7645-4084-X
- Macromedia Flash MX 2004 For Dummies 0-7645-4358-X
- Photoshop 7 All-in-One Desk Reference For Dummies 0-7645-1667-1
- Photoshop cs Timesaving Techniques For Dummies 0-7645-6782-9
- PHP 5 For Dummies 0-7645-4166-8
- PowerPoint 2003 For Dummies 0-7645-3908-6
- QuarkXPress 6 For Dummies 0-7645-2593-X

NETWORKING, SECURITY, PROGRAMMING & DATABASES

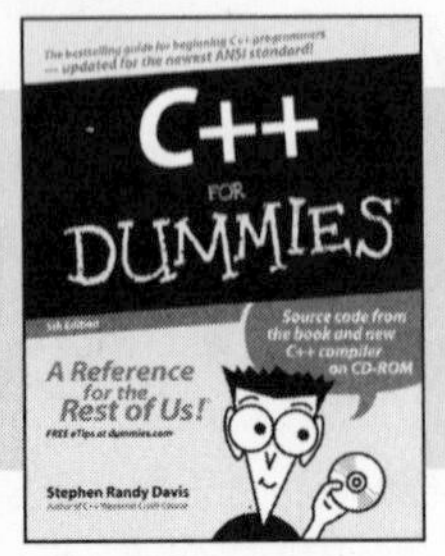

0-7645-6852-3

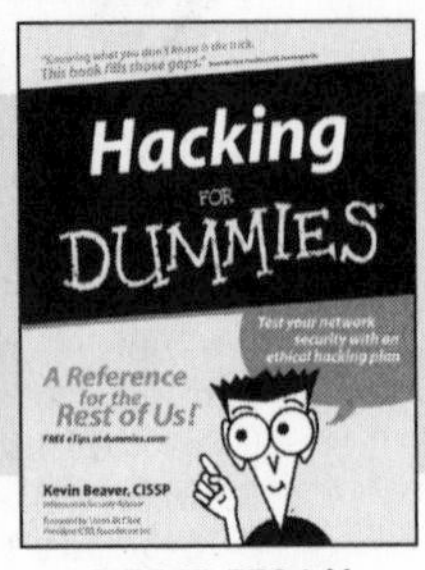

0-7645-5784-X

Also available:

- A+ Certification For Dummies 0-7645-4187-0
- Access 2003 All-in-One Desk Reference For Dummies 0-7645-3988-4
- Beginning Programming For Dummies 0-7645-4997-9
- C For Dummies 0-7645-7068-4
- Firewalls For Dummies 0-7645-4048-3
- Home Networking For Dummies 0-7645-42796
- Network Security For Dummies 0-7645-1679-5
- Networking For Dummies 0-7645-1677-9
- TCP/IP For Dummies 0-7645-1760-0
- VBA For Dummies 0-7645-3989-2
- Wireless All In-One Desk Reference For Dummies 0-7645-7496-5
- Wireless Home Networking For Dummies 0-7645-3910-8